25-44

PUBLISHERS' NOTE

The series in which this title appears was introduced by the publishers in 1957 and is under the general editorship of Dr. M. G. Kendall. It is intended to fill a need which has been evident for some time and is likely to grow: the need for some form of publication at moderate cost which will make accessible to a group of readers specialized studies in statistics or special courses on particular statistical topics. There are numerous cases where, for example, a monograph on some newly developed field would be very useful, but the subject has not reached the stage where a comprehensive book is possible; or again, where a course of study is desired in a domain not covered by text-books but where an exhaustive treatment, even if possible, would be expensive and perhaps too elaborate for the readers' needs.

Considerable attention has been given to the problem of producing these books speedily and economically. Appearing in a cover the design of which will be standard, the contents of each volume will follow a simple, straightforward layout, the text production method adopted being suited to the complexity or otherwise of the subject.

The publishers will be interested in approaches from any authors who have work of importance suitable for the series.

CHARLES GRIFFIN & CO. LTD.

OTHER BOOKS ON MATHEMATICS AND STATISTICS

GREEN'S FUNCTION METHODS IN PROBABILITY THEORY

JULIAN KEILSON

Senior Scientist, Applied Research Laboratory, Sylvania
Electronic Systems, General Telephone & Electronics, Inc.

BEING NUMBER SEVENTEEN OF
GRIFFIN'S STATISTICAL
MONOGRAPHS & COURSES

EDITED BY
M. G. KENDALL, M.A., Sc.D.

1965
HAFNER PUBLISHING COMPANY
NEW YORK

CHARLES GRIFFIN & COMPANY LIMITED
42 DRURY LANE, LONDON, W.C.2

First published in 1965

DEDICATION

To PAULA

Printed in Great Britain by Latimer Trend & Co Ltd Whitstable

PREFACE

The domain of interest of this book is the family of Markov processes which are homogeneous in space and time, and the family of related inhomogeneous Markov processes induced by boundaries. A simple example in discrete time of a homogeneous Markov process is the sequence $S_k = \sum_1^k \xi_i$ of sums of independent identically distributed random variables. Other examples in continuous time are provided by the homogeneous diffusion process, the compound Poisson process, and more generally the class of processes with independent increments. Closely associated with these homogeneous processes, but somewhat more involved in structure, are the Markov processes obtained when one or two boundaries, absorbing or impenetrable, are introduced. These processes have many important applications and have featured prominently over the last fifteen years in the literature of queues, dams, inventories, and sequential sampling.

Our object is two-fold. First we wish to describe the homogeneous processes on the lattice and continuum for discrete and continuous time, their interrelation, the structure of their transition functions, and the character and structure of certain associated ergodic Green's functions studied in potential theory. Emphasis is given to asymptotic behaviour in space and time and to the analytical tools of conjugate transformation, saddlepoint integration, and related methods in the complex plane employed to exhibit this behaviour. Second, we wish to show how the transition functions and ergodic Green's functions for the homogeneous processes enter into the description of the related bounded processes as natural entities, and permit an asymptotic discussion of their transient and ergodic behaviour.

The presentation is largely self-contained. It is assumed that the reader will have a background in the elements of probability theory and of stochastic processes, and will also have a familiarity with functions of a complex variable. Applications are discussed in detail.

This book had its origins in a series of lectures at the University of Birmingham in the Spring of 1963. It would not have been written without the encouragement and support of Professor H.E. Daniels, to whom I extend my sincere appreciation. I also wish to express my gratitude to my colleague D.M.G. Wishart for devoting much of his energy over the past three years to the development of these ideas and for reading the manuscript. I thank M. Zakai for his generous attention and many helpful suggestions, N.U. Prabhu for several useful comments and references, my associates A. Kooharian and S.F. Neustadter for their continuing interest, and Mrs. Barbara Pszczolkowski for typing the manuscript and helping proofread it. I thank W.L. Smith for calling my attention to a serious gap in an early proof.

Finally, I would like to convey my appreciation to Sylvania Electronic Systems, and in particular to H. Lehne, L.S. Sheingold and J.E. Storer, for supporting this undertaking.

1st June 1965 J.K.

CONTENTS

CHAPTER I

INTRODUCTION AND SUMMARY

I.0 In this introductory chapter, an attempt will be made to outline the scope of the book and its motivation. The topics treated will be described briefly, and the character of the methods and principal results obtained indicated. Introductory chapters of this kind are inevitably fragmentary, and many of the terms employed will only be properly defined within the body of the book, which commences with Chapter II.

I.1 Homogeneous Markov processes modified by boundaries

In a great variety of practical disciplines, stochastic processes are employed to model the behaviour of a system governed by random events. The natural role of such processes in the study of capacity and delay in communication systems, dams, inventories, collective risk, population growth, epidemics, biophysics, reliability, etc., has been well documented.† A family of Markov processes frequently encountered in such studies may be described as spatially and temporally homogeneous Markov processes, modified by the presence of one or two boundaries.

A simple example arises in the study of dams. The amount of water present in a reservoir is a function of time $X(t)$ whose variation is governed by the following laws. In any time interval $(t, t+dt)$ there is a probability νdt of a finite increment to the reservoir. The increments received, ξ_1, ξ_2, etc., are independent random variables having a prescribed distribution $A(x) = P\{\xi \leqslant x\}$. Water leaves the reservoir at a constant rate of efflux, $v < 0$. The reservoir has a given capacity L, and any water in excess of this capacity is lost.

If the effect of the finite capacity and of the possibility of emptying could be ignored, the amount of water present might be represented by the process $X^*(t) = \sum_1^{K(t)} \xi_i + vt$ where $K(t)$ is the number of rainfalls occurring in $(0,t)$. The process $X^*(t)$ is spatially and temporally homo-

†See, for example, Bartlett (1955) and Bharucha-Reid (1960).

geneous in the sense that for any $0 \leqslant t_1 < t_2$ the change in level $X^*(t_2) - X^*(t_1)$ is a random variable whose distribution is independent of the level at t_1 (as would be true if the capacity were infinite and emptying could not occur) and of t_1 itself (the increment distribution is independent of time). The process $X^*(t)$ is unbounded and assumes all levels x in $(-\infty,\infty)$. Because of the finite capacity L, and the intrinsic non-negative character of the amount of water present, the actual process $X(t)$ is confined to the finite interval $0 \leqslant x \leqslant L$. The spatial homogeneity may be said to be modified by boundaries at $x = 0$ and $x = L$.

All of the processes of ultimate concern to us, whether processes in discrete or continuous time; lattice processes or continuum processes; purely discontinuous, continuous, or of mixed character; whether on a finite or semi-infinite interval – all will have this bounded homogeneous character. (The precise sense in which these and other terms of this introductory chapter are employed will be given in the main body of the text.) Many other examples are given in the first section of Chapter V.

In the description of such processes, certain random variables have been of recurring interest:

(1) *Passage times.* If at $t = 0$ it is known that the process is in the state $X_0 = y$, the passage time $\mathcal{T}_y$ at which a boundary is first reached or exceeded is important. The distribution $S_y(x) = P\{\mathcal{T}_y \leqslant x\}$ is desired. For our reservoir example, the time at which the reservoir first empties, and the time until the first overflow would be of interest.

(2) *Which of two boundaries is first reached.* When the process is confined to a finite interval $[0,L]$ by two boundaries, the probability R_y that a particular boundary, say $x = L$, will be reached before $x = 0$, given that the process starts at y, is of frequent interest. Such problems are sometimes called gambler's ruin problems.

(3) *The value at ergodicity.* If the process is known to be ergodic, i.e., to reach a unique non-trivial distribution after sufficient time independent of the initial state of the system, its ergodic distribution is important. For the reservoir process one is interested in the distribution of $X(\infty)$.

(4) *The time between certain recurrent events.* When the process is ergodic, a class of recurrent events may be of particular interest. If the distribution of the process subsequent to such an event is known, the Markov character of the process implies that the time between successive events will constitute a sequence of independent identically distributed random variables. For the reservoir process,

one such class of events is the set of times at which the dam empties, i.e., the depletion times. For the same process, empty periods alternate with non-empty periods, and the durations of non-empty periods are of interest.

(5) *The greatest value attained between depletions.* Let $x = 0$ be the smallest state assumed and let entrance into the state $x = 0$ be recurrent. The maximum value assumed by the process between successive depletions is of frequent interest. This random variable is sometimes called the extreme.

As we will see, these random variables and their distributions are closely related. The relationships may be employed to exhibit the explicit distributions in simple cases.

I.2 Methods of solution in the complex plane

The methods that have been employed to obtain the distributions desired have generally centred about considerations in the complex plane. For the dam process of Section 1, for example, when the reservoir has infinite capacity, the durations τ of the periods when water is present have a distribution $F(x) = P\{\tau \leqslant x\}$ whose Laplace-Stieltjes transform $\phi^*(s)$ may be shown (Kendall, 1951) to satisfy the functional equation

$$\phi^*(s) = \alpha^*(s+\nu-\nu\phi^*(s)). \tag{2.1}$$

Here ν is the mean rate at which increments occur and $\alpha^*(s)$ is the Laplace-Stieltjes transform of the increment distribution. A characterization of the form (2.1) as the final product of an involved calculation leaves much to be desired.

A second example is the process on the half-line $[0,\infty)$,

$$X_k = \max[0, X_{k-1} + \xi_k] \tag{2.2}$$

introduced to queuing theory by Lindley (1952). The random variables ξ_k are assumed to be independent and to have a common distribution $A(x)$. The process (2.2) may be regarded as a modification of the homogeneous process $X_k^* = X_{k-1}^* + \xi_k$ by a boundary at $x = 0$ which prevents the homogeneous process from attaining a negative value. If the mean increment $\mu = \int x\, dA(x)$ is negative, Lindley has shown that the process is ergodic and that $F_k(x)$, the distribution of X_k, approaches a unique ergodic distribution $F_\infty(x)$ as $k \to \infty$. This ergodic distribution is characterized by a Wiener-Hopf type integral equation (V.12), seeking the factorization of a prescribed kernel function in the complex plane. A useful discussion of the ergodic distribution $F_\infty(x)$ via this characterization hinges on one's ability to find the zeros of the kernel in one half-plane or the other. The zeros required are not readily found, except in simple cases, and a solution expressed in terms of unknown zeros is largely formal and of limited practical value.

Much of the literature on bounded homogeneous processes is handicapped by the inadequacy of such formulation, and this limitation has been widely recognized. In some instances, alternative descriptions in the real domain have been exhibited. For example, the distribution $F(x)$ characterized by (2.1) has been exhibited by Kendall (1957) heuristically and by Prabhu (1960) and Takács (1961) by combinatorial methods.† A general approach to the practical analysis of bounded homogeneous processes avoiding such limitation has not been available.

The Green's function methods described in this book are an attempt in this direction.

I.3 Green's functions

Green's function methods play a basic role in the analysis of linear equations in the presence of boundaries. Consider for example, the spatially homogeneous diffusion equation

$$\frac{\partial}{\partial t} f(x,t) = \frac{1}{2}\frac{\partial^2}{\partial x^2} f(x,t), \tag{3.1}$$

governing a particle density function $f(x,t)$. When the diffusion takes place in the presence of absorbing boundaries at $x = a$ and $x = b$, the function $f(x,t)$ must satisfy the boundary conditions $f(a,t) = 0$, and $f(b,t) = 0$. An initial density $f(x,0) = f_0(x)$ is specified. When boundaries are absent, the natural solution of (3.1) for an initial concentration at $x = 0$, i.e., for $f(x,0) = \delta(x)$, is provided by the familiar transition probability density

$$g(x,t) = (2\pi t)^{-\frac{1}{2}} \exp(-x^2/2t). \tag{3.2}$$

This solution, in the absence of finite boundaries for initial concentration at zero, is called a Green's function for equation (3.1). The system of equations for the absorption problem may then be shown to have a solution of the form

$$f(x,t) = \int_a^b f_0(x')\, g(x-x', t)\, dx' \tag{3.3}$$

$$- \int_0^t l_a(t')\, g(x-a, t-t')\, dt' - \int_0^t l_b(t')\, g(x-b, t-t')\, dt'$$

where $l_a(t)$ and $l_b(t)$ are unknown functions associated with the absorbing boundaries. The boundary conditions $f(a,t) = f(b,t) = 0$ employed

† The Spitzer identity should also be noted.

with (3.3) give a pair of simultaneous integral equations for $I_a(t)$ and $I_b(t)$, of convolution type. If Laplace transformation is employed, the solution in the transform plane is immediate. The essence of the situation is that the introduction of the Green's function acts to convert the problem from a fundamentally complex equation involving all states of the interval, to a much simpler set of equations associated only with the boundaries.

A similar reduction in complexity takes place for many bounded homogeneous processes when Green's functions are employed. A process of this type is governed by a continuity equation or set of continuity equations interconnecting the time-dependent state probabilities. These continuity equations for the bounded processes may be regarded as equivalent to the continuity equation for the homogeneous process present when the bounds are removed, modified by the presence of a set of auxiliary source terms representing the effect of the boundaries.

Consider, for example, the symmetric random walk $N(t)$ on the lattice in continuous time, modified by a reflecting boundary at the state $n = 0$. Let $N(0) = n_0$, and let $p_n(t)$ be the probability that $N(t) = n$. Then if νdt is the probability of a step in time dt, the continuity equations for $p_n(t)$ are given by

$$\frac{dp_0}{dt} = -\frac{1}{2}\nu p_0(t) + \frac{1}{2}\nu p_1(t) \tag{3.4a}$$

and

$$\frac{dp_n}{dt} = -\nu p_n(t) + \frac{1}{2}\nu p_{n-1}(t) + \frac{1}{2}\nu p_{n+1}(t), \quad n \geqslant 1. \tag{3.4b}$$

The initial distribution is $p_n(0) = \delta_{n,n_0}$. This system of equations is equivalent to the system

$$\frac{dp_n}{dt} = -\nu p_n(t) + \frac{1}{2}\nu p_{n-1}(t) + \frac{1}{2}\nu p_{n+1}(t) + c_n(t), \quad -\infty < n < \infty \tag{3.5a}$$

valid for *all* n, and

$$c_n(t) = -\delta_{n,-1}\left\{\frac{1}{2}\nu p_0(t)\right\} + \delta_{n,0}\left\{\frac{1}{2}\nu p_0(t)\right\} \tag{3.5b}$$

where $\delta_{m,n}$ is the Kronecker delta symbol. The equivalence may be seen heuristically in the following way. It is easy to show that the set of negative states with $n < 0$ remains empty for all time. For from (3.5a) $(d/dt)\sum_{-\infty}^{-1} p_m(t) = -\frac{1}{2}\nu p_{-1}(t) + \{\frac{1}{2}\nu p_0(t) + c_{-1}(t)\}$. The net rate of arrival to the set is $\frac{1}{2}\nu p_0(t) + c_{-1}(t)$, which from (3.5b) has the value zero for all time t. Hence the set, which is initially empty, must remain empty, and in particular $p_{-1}(t) = 0$. From (3.5a) for $n = 0$ and

(3.5b) we then find (3.4a), and the two systems are identical.

The Green's function natural to the system is the solution of the homogeneous system (3.5a) with $c_n(t) = 0$ and $p_n(0) = \delta_{n,0}$, i.e., the transition probability for the homogeneous process. By a generating function technique (II.9), we find that the Green's function is

$$g_n(t) = e^{-\nu t} I_n(\nu t) \tag{3.6}$$

where $I_n(t)$ is the modified Bessel function of order n. Hence the solution of (3.5a) is

$$p_n(t) = g_{n-n_0}(t) + c_{-1}(t) * g_{n+1}(t) + c_0(t) * g_n(t) \tag{3.7}$$

where the asterisk denotes convolution in time. From (3.5b) we see that $c_{-1}(t) = -c_0(t)$ and $c_0(t) = \frac{1}{2}\nu p_0(t)$. Hence, from (3.7), the system of equations (3.4) is reduced to a single integral equation

$$c_0(t) = \frac{1}{2}\nu g_{-n_0}(t) + \frac{1}{2}\nu c_0(t) * [g_0(t) - g_1(t)] \tag{3.8}$$

which may be solved by Laplace transformation. A faster and more intuitive solution is obtained in Chapter IV by an image method which demonstrates vividly the basic role played by the Green's function in the solution.

By the Green distribution for a homogeneous process modified by boundaries, we will always mean the time-dependent transition distribution for the underlying homogeneous process. In the potential theory of (lattice) random walks in one or more dimensions (Spitzer, 1964), the "Green function" plays a basic role. This entity in a more general form will appear for us with the name "ergodic Green's function", and will soon be discussed.

I.4 Homogeneous Markov processes

Spatially and temporally homogeneous processes, modified by boundaries, appear in applications in a great many forms. The underlying homogeneous processes in discrete and continuous time may all be related to the basic homogeneous process in discrete time,

$$X_k = \xi_1 + \xi_2 + \dots + \xi_k , \tag{4.1}$$

which consists of successive partial sums of a sequence $\xi_1, \xi_2, \dots,$ etc., of independent, identically distributed random increments. The time-dependent transition distribution or Green distribution for this process, $G_k(x) = P\{X_k \leqslant x\}$, is the k-fold convolution (II.3)

$$G_k(x) = A^{(k)}(x) \tag{4.2}$$

of the increment distribution $A(x)$. Closely related to these basic homogeneous processes in discrete time are the homogeneous jump processes in continuous time. With any process (4.1) governed by the increment distribution $A(x)$, there may be associated a jump process (II.2, II.3)

$$X_J(t) = \sum_{1}^{K(t)} \xi_k . \tag{4.3}$$

In any interval of length dt, there is probability $\nu\, dt$ that an increment will occur. Thus $K(t)$ is an auxiliary Poisson process of intensity ν giving the number of increments that occur in the interval $(0,t)$. The probability of k increments in $(0,t)$ is $e^{-\nu t}(\nu t)^k/k!$. Hence from (4.2) the transition distribution or Green distribution, $G(x,t) = P\{X(t) \leqslant x\}$ is

$$G(x,t) = \sum_{0}^{\infty} e^{-\nu t} \frac{(\nu t)^k}{k!} A^{(k)}(x), \tag{4.4}$$

where $A^{(o)}(x) = U(x)$, the degenerate distribution with all support at $x = 0$. The homogeneous jump processes are also known as compound Poisson processes. The processes (4.1) and (4.3), when specialized to the lattice, have distinctive properties and notation (II.4).

A trivial extension of (4.3) of great importance in practice adds a linear growth term vt to the process, giving

$$X_V(t) = \sum_{1}^{K(t)} \xi_k + vt \tag{4.5}$$

so that $G_V(x,t) = G_0(x-vt,t)$. The triviality of the extension disappears in the presence of boundaries, and the ergodic distributions have a distinct character. It will be convenient to refer to such processes as homogeneous drift processes.

The general homogeneous process $X(t)$ in continuous time has the property that the increment $X(t_2) - X(t_1)$ in any time interval (t_1,t_2) is independent of that in any other non-overlapping time interval. Such processes are often called processes of independent increments. The Green distribution is then infinitely divisible (II.3d), and conversely with any infinitely divisible distribution, there may be associated a homogeneous process in continuous time. Because of the infinite divisibility of the Green distribution, any homogeneous process in continuous time may be shown to be the limit of some sequence of jump processes $X_j(t)$ as a simple consequence of DeFinetti's Theorem (II.7).

Such a representation provides a useful tool for studying the drift process and other general homogeneous processes in continuous time when modified by boundaries. The most important example is provided by the Wiener-Lévy process $X_D(t)$ for which the Green density $g(x,t) = \frac{d}{dx} P\{X_D(t) \leqslant x\}$ is given by

$$g_D(x,t) = (4\pi Dt)^{-\frac{1}{2}} \exp\{-(x-vt)^2/4Dt\}. \tag{4.6}$$

It will often be convenient to discuss a homogeneous process $X(t)$ having a jump component, drift component and Wiener-Lévy component, i.e.,

$$X(t) = X_J(t) + vt + X_D(t). \tag{4.7}$$

Chapter II is devoted to the study of the structure of these homogeneous processes, their equations of continuity, and their Green's functions. The basic properties of characteristic functions, two-sided generating functions, and associated inversion theorems are reviewed. Some simple Green's functions are exhibited.

I.5 Conjugate transformation and the Central Limit Theorem

For any probability distribution $F(x)$, a family of conjugate probability distributions

$$F_V(x) = \int_{-\infty}^{x} e^{-Vx'} dF(x') \Big/ \int_{-\infty}^{\infty} e^{-Vx'} dF(x') \tag{5.1}$$

may be defined whenever the integration in the denominator converges for real non-zero values of V. These conjugate distributions are associated with Khinchin. The denominator is the value of the characteristic function $\phi(z) = \int_{-\infty}^{\infty} e^{izx} dF(x)$ at $Z = iV$, and a non-trivial conjugate family of distributions is available whenever the characteristic function has a convergence strip (III.1) of finite width. The conjugate distributions provide a basic tool for the study of asymptotic behaviour and convergence, and will play a prominent role throughout the book.

The characteristic function $\phi(z)$ of a possible distribution $F(x)$ is plotted in Fig.I.1 at values $z = iV$ on the imaginary axis inside the convergence strip. The convexity of the function $\phi(iV)$ is readily demonstrated for any distribution. The convergence strip is bounded by its upper and lower boundaries V_{min} and V_{max}. For each value V of the convergence interval (V_{min}, V_{max}) there will be a conjugate distribution $F_V(x)$ defined as above. Three important values of V are indicated in the figure. The value $V = 0$ corresponds to the original distribution $F_0(x) = F(x)$. The value V_0 is located at the minimum of $\phi(iV)$, and the value V_1 is that value other than zero at which $\phi(iV) = 1$. Most of the convergence strips† arising in applications will have such values V_0 and V_1. The corresponding conjugate distributions $F_{V_0}(x)$ and $F_{V_1}(x)$ play an important role in the asymptotic studies. Of particular interest are the associated numbers $\phi(iV_0)$, and $\mu_{V_1} = \int x\, dF_{V_1}(x)$, the first moment of the conjugate distribution for V_1.

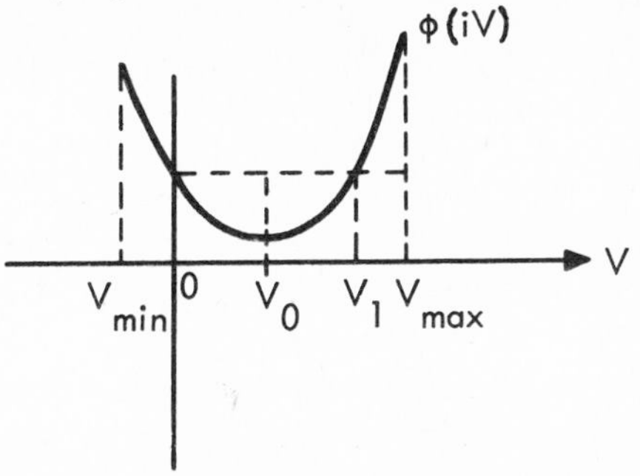

Fig. I.1

The Green distributions, $G_k(x) = A^{(k)}(x)$ for discrete time and $G(x,t) = G^{(\nu t)}(x,\nu^{-1})$ for continuous time have an asymptotic behaviour with k obtained from the theory surrounding the Central Limit Theorem. As will be seen, asymptotic behaviour for the bounded homogeneous processes is closely related to questions of "large deviations", e.g., to the behaviour of $A^{(k)}(x)$ for large k when x falls outside the region surrounding the mean $k\mu$ in which the central limit approximation may be expected to be valid. Chapter III is devoted to a simplified account of conjugate transformation, the Central Limit Theorem, and large deviations for the variety of Green distributions encountered in the study of homogeneous processes.

A very efficient method of treating large deviations, introduced by Daniels (1954), is the saddlepoint approximation. The saddlepoint approximation has a simple interpretation in terms of the conjugate family of distributions. If one works with the identity $F^{(k)}(x) = \{\int_{-\infty}^{x} e^{Vx'} dF_V^{(k)}(x')\}\,\phi^k(iV)$, (cf. III.2), one may choose for a given value of x that distribution $F_V(x)$ for which $F_V^{(k)}(x)$ has mean $k\mu_V = x$ and then employ the Central Limit Theorem. In this way one employs the Central Limit Theorem in an optimum manner. The use of conjugate transformation to derive the saddlepoint approximation has the advantage of permitting discussion of distributions which are not absolutely continuous. If we designate the optimum value of V by θ, the saddle-

† (for distributions with both positive and negative support).

point approximation for a local density is (III.4)

$$f_0^{(N)}(x) \approx \phi_0^N(i\theta)\, e^{\theta x}\,(2\pi N\sigma_\theta^2)^{-\frac{1}{2}} . \tag{5.2}$$

The saddlepoint approximation gives rise to a coarse but simple criterion for estimating the range of values (x,N) for which the local central limit approximation (III.5.1)

$$f_0^{(N)}(x) \approx (2\pi N\sigma_0^2)^{-\frac{1}{2}} \exp\{-(x-N\mu_0)^2/(2N\sigma_0^2)\} \tag{5.3}$$

may be expected to be valid. If the departure of $f_0(x)$ from normality is not excessive, one must satisfy the two conditions

$$\text{(a)} \quad N \gg 1 \qquad \text{and} \qquad \text{(b)} \quad \frac{|x - N\mu_0|}{N^{2/3}\,\sigma_0} \ll 1 . \tag{5.4}$$

A family of Gaussian approximations in different "zones of convergence" is available in the same way.

The asymptotic expansion of $f_0^{(N)}(x)$ in powers of N^{-1} associated with the saddlepoint approximation is discussed in Section III.5, and the first correction term exhibited.

The analogues of these results for lattice processes and processes in continuous time are dealt with in Sections III.6, III.7 and III.8.

I.6 Skip-free processes

A process is said to be skip-free in the negative direction if a sample cannot pass from a value x_1 to any lower value $x_2 < x_1$ without passing through all the intervening states. Lattice processes with only unit steps permitted in the negative direction, drift processes such as (4.5) with negative drift rate v and positive increments, and the homogeneous diffusion processes are all skip-free negative.

The lattice processes with unit increments only in both directions (Bernoulli increments) have a particularly simple structure. For such a process in discrete or continuous time, modified by either a reflecting or absorbing boundary at $n = 0$, the transient distributions can be exhibited simply in terms of the Green distribution without resort to generating functions or Laplace transformation (IV.2). The method of images and use of the conjugate family of distributions suffice.

When the lattice process $N(t)$ is skip-free in only one direction, a generalization of the method of images is available (IV.3). The passage time τ from some initial positive state n_0 to the state $n = 0$ has a simple probability density $s_{n_0}(\tau)$ given by

$$s_{n_0}(\tau) = \frac{n_0}{\tau}\, g_{-n_0}(\tau) \tag{6.1}$$

where $g_n(t) = P\{N(t) = n | N(0) = 0\}$ is the Green probability for the process. The passage time probabilities for the discrete time case have a similar form.

Similarly, for any process $X(t)$ on the continuum which is skip-free negative, the passage time τ from any initial positive state y to the state 0 has a probability density

$$s_y(\tau) = \frac{y}{\tau} g(-y,\tau) \tag{6.2}$$

where $g(x,t) = \dfrac{d}{dx} P\{X(t) \leqslant x | X(0) = 0\}$ is the local Green density, which may only exist in a generalized sense.

Because, as we have indicated in Section 5, the asymptotic behaviour of the Green's functions for large time is readily available, the behaviour of $s_{n_0}(\tau)$ and $s_y(\tau)$ for large τ is equally accessible from (6.1) and (6.2). The character of the diffusion approximation to the passage time density and its range of validity may then be understood in terms of the zone of convergence appropriate.

A simple extension of the result (6.2) to a class of processes which are not skip-free is given in Section VII.4.

I.7 The ergodic Green's functions and the structure of the ergodic distributions

Consider a homogeneous process in discrete time modified by boundaries, which we denote by X_k. Let the process be governed by a single-step transition distribution $A(x', x) = P\{X_{k+1} \leqslant x | X_k = x'\}$ assumed for the moment to be absolutely continuous in x for all x', with probability density $a(x', x) = (d/dx)\, A(x', x)$. There will then be for $k > 1$ a sequence of densities $f_k(x) = (d/dx) P\{X_k \leqslant x\}$ describing X_k, governed by the continuity equation

$$f_{k+1}(x) = \int f_k(x')\, a(x', x)\, dx'. \tag{7.1}$$

We may employ the homogeneity away from the boundaries by writing $a(x', x) = a(x - x') + d(x', x)$ where $a(x)$ is the increment density for the underlying homogeneous process and $d(x', x) = a(x', x) - a(x - x')$. It may then be shown (V.5, V.11) that if the process X_k is ergodic, the limiting ergodic distribution $f_\infty(x)$ has the form

$$f_\infty(x) = c_\infty(x) + \int_{-\infty}^{\infty} h(x - x')\, c_\infty(x')\, dx' = \int_{-\infty}^{\infty} g(x - x')\, c_\infty(x')\, dx' \tag{7.2}$$

where $h(x)$ is the extended renewal density, defined by

$$h(x) = \sum_1^{\infty} a^{(n)}(x); \qquad g(x) = \delta(x) + h(x). \tag{7.3}$$

The function $g(x)$ is called the ergodic Green density. The function $c_\infty(x)$ is a compensation density related to $f_\infty(x)$ by

$$c_\infty(x) = \int_{-\infty}^{\infty} f_\infty(x')\,d(x',x)\,dx'. \tag{7.4}$$

Equation (7.2) states that the ergodic distribution is the convolution of the compensation function and the ergodic Green density. The usefulness of this relation will soon be apparent.

The series (7.3) for $h(x)$ converges almost everywhere when $a(x)$ has a non-zero first moment (V.6.1, Notes 7 and 8).

Direct analogues of (7.1), (7.2), (7.3) and (7.4) are available for an ergodic process N_k on the lattice, where for example (7.2) takes the form $p_n^{(\infty)} = P\{N_\infty = n\} = \sum_{-\infty}^{\infty} c_{n'}^{(\infty)} g_{n-n'}$ (V.7).

For more general $A(x)$ and $A(x',x)$ one may work with comparable relations involving the ergodic distribution $F_\infty(x)$, the extended renewal function

$$H(x) = \sum_{1}^{\infty} \{A^{(k)}(x) - A^{(k)}(0)\} \tag{7.5}$$

and an ergodic compensation function

$$C_\infty(x) = \int \{A(x',x) - A(x-x')\}\,dF_\infty(x'). \tag{7.5'}$$

An important example of the kind of process considered is the process X_k of (2.2). Here it may be seen that $A(x',x) = A(x-x')U(x)$ for $x' \geqslant 0$. The boundary at $x = 0$ which acts to prevent negative values may be called a retaining† boundary. A second type of process of interest modifies the homogeneous process by replacement at a new value T whenever the homogeneous process goes negative. The new value may either be a constant or may itself be a random variable of prescribed distribution. Such processes may be called replacement † processes. Examples arising in practice are given in Section V.1.

A similar class of bounded processes is of interest for processes in continuous time. Consider the jump process $X(t)$ whose transitions are governed by a Poisson process of frequency ν and a single step transition density $a(x',x)$. If the underlying homogeneous process has increment density $a(x)$, one has in place of (7.2), (7.3) and (7.4),

$$f(x,\infty) = \int_{-\infty}^{\infty} g_c(x-x')\,c(x',\infty)\,dx', \tag{7.6}$$

† In the literature the retaining boundary is often described as impenetrable, and the replacement process as a return process.

where

$$\mathrm{g}_c(x) = \int_0^\infty g(x,t)\,dt\,, \tag{7.7}$$

and $c(x,\infty)$, the limiting form of a compensation function $c(x,t)$, is

$$c(x,\infty) = \nu \int_{-\infty}^{\infty} f(x',\infty)\{a(x',x) - a(x-x')\}\,dx'. \tag{7.8}$$

The function $g(x,t)$ is the transition density obtained from (4.4), by differentiation. The function $\mathrm{g}_c(x)$ of (7.7) is called the ergodic Green density for continuous time. The equations (7.6), (7.7) and (7.8) are in a sense trivial since it may be quickly shown that $f(x,\infty)$ and $f_\infty(x)$ of (7.2) are identical. These equations, however, permit one to discuss bounded processes whose underlying homogeneous processes are not of the simple jump type, and in particular the drift processes. For a process of type (4.7), the basic condition that the integral (7.7) converge is that $\int x\,g(x,1)\,dx \neq 0$ (Theorem V.9.4).

I.8 Simple properties of the ergodic Green's functions

Simple properties of the ergodic Green's functions provide a key to the discussion of the ergodic distributions. Of greatest practical interest is the information about asymptotic behaviour available.

The extended renewal density $h(x)$ appearing in (7.2) and (7.3) has the following characteristic properties ($\int x\,a(x)\,dx \neq 0$):

(1) If $a(x)$ is bounded for all x on $(-\infty,\infty)$, then (Theorem V.6.1) $h(x)$ is continuous wherever $a(x)$ is, and is bounded on $(-\infty,\infty)$. If $\mu = \int x\,a(x)\,dx < 0$, and $a(x) \to 0$ as $x \to \pm\infty$,

$$\lim_{x\to-\infty} h(x) = |\mu|^{-1}, \qquad \lim_{x\to+\infty} h(x) = 0. \tag{8.1}$$

(2) When $\mu < 0$ and the negative increments described by $a(x)$ are exponentially distributed, the relation

$$h(x) = |\mu|^{-1} \tag{8.2}$$

is exact for all $x < 0$ (Theorem V.8.1). This structural property has important consequences for ergodic processes on finite intervals.

(3) When (a) $\mu < 0$; (b) at a point $V_1 < 0$ in the interior of the convergence interval, $\alpha(iV_1) = 1$; (c) $a(x)e^{-V_1 x} \to 0$ as $x \to \infty$; then

$$h(x) \sim \mu_{V_1}^{-1} e^{V_1 x} \quad \text{as } x \to \infty\,, \tag{8.3}$$

where μ_{V_1} is the first moment of $a_{V_1}(x)$ (Theorem V.10.1). When the family of conjugate densities is sufficiently large, an asymptotic expansion of $h(x)$ of which (8.3) is the first term is available (Theorem V.10.2).

In place of the extended renewal function $H(x)$ of (7.5), it is often desirable to work with one of the related functions $H^+(x) = \sum_1^\infty A^{(k)}(x)$ and $H^-(x) = \sum_1^\infty \{A^{(k)}(x) - 1\}$. The convergence of these functions may be inferred from the behaviour of the associated power series

$$H^+(w,x) = \sum_1^\infty w^k A^{(k)}(x); \quad H^-(w,x) = \sum_1^\infty w^k\{A^{(k)}(x) - 1\}, \tag{8.4}$$

discussed in Section V.6.1, where the following Theorem is demonstrated.

Theorem V.6.3. For any increment distribution $A(x)$, let $\alpha(iV_+^*)$ be the greatest lower bound of the set of values $\{\alpha(iV)\colon\ 0 \leqslant V < V_{\max}\}$, in the convergence strip; let $\alpha(iV_-^*)$ be the greatest lower bound of $\{\alpha(iV)\colon\ V_{\min} < V \leqslant 0\}$. Then the series $H^+(w,x) = \sum_1^\infty w^n A^{(n)}(x)$ has the radius of convergence $\rho_+ \geqslant [\alpha(iV_+^*)]^{-1}$; the series $H^-(w,x) = \sum_1^\infty w^n\{A^{(n)}(x) - 1\}$ has the radius of convergence $\rho_- \geqslant [\alpha(iV_-^*)]^{-1}$.

From this Theorem it follows that $H^+(x)$ converges when μ, the first moment of $A(x)$, is positive and $V_{\max} > 0$. Similarly, $H^-(x)$ converges when $\mu < 0$ and $V_{\min} < 0$. It also follows at once from the theorem, that if $A(x)$ has only positive support, $\sum_1^\infty w^k A^{(k)}(x)$ is an *entire* function of w for every x.

The set of properties (1), (2) and (3) above for $h(x)$ may also be placed in evidence for lattice processes (Theorems V.7.1, V.8.3, V.8.3$'$ and V.8.4), and for the ergodic Green's functions for the homogeneous processes in continuous time.

Consider in particular the process $X(t) = X_J(t) + vt + X_D(t)$ of (4.7). The transition distribution $G(x,t)$ is absolutely continuous whenever the Wiener-Lévy parameter D of (4.6) is positive. It may then be shown that the ergodic Green density $g_c(x) = \int_0^\infty g(x,t)\,dt$ of (4.7) converges (Theorem V.9.4) when, for $\mu = \int x\,dA(x)$

$$\mu(1) = \int_0^\infty x\,g(x,1)\,dx = \nu\mu + v \neq 0, \tag{8.5}$$

and that $g_c(x)$ has the following properties:

(1) $g_c(x)$ is continuous for all x. Its asymptotic behaviour is given by $\lim_{x\to\infty} g_c(x) = 0$ and $\lim_{x\to-\infty} g_c(x) = |\mu(1)|^{-1}$ if $\mu(1) < 0$ (Theorem V.9.4).

(2) If the increment distribution $A(x)$ governing $X_J(t)$ has no nega-

tive support, the process $X(t)$ is skip-free in the negative direction and $g_c(x) = |\mu(1)|^{-1}$ for all $x \leqslant 0$, when $\mu(1) < 0$ (Theorem V.9.5).

(3) When $\mu(1) < 0$ and the convergence strip for the characteristic function of $G(x,1)$ contains the point V_1 of Fig. I.1, then $g_c(x)$ has the asymptotic representation $g_c(x) \sim \{\mu_{V_1}(1)\}^{-1} e^{V_1 x}$ as $x \to \infty$, where $\mu_{V_1}(1)$ is the first moment of the conjugate distribution $G_{V_1}(x,1)$ (Theorem V.10.3).

(4) When $D = 0$, $X(t)$ is a homogeneous drift process. The ergodic Green measure $G_c(x) = \int_0^\infty [G(x,t) - G(0,t)]\, dt$ corresponding to (7.7) is again absolutely continuous, and continuous at all $x \neq 0$, when $\mu(1) = \int x\, dG(x,1) = \nu\mu + v \neq 0$ (Theorem V.9.1). Properties (1), (2) and (3) remain valid apart from the discontinuity of $g_c(x)$ at $x = 0$ (Theorems V.9.2, V.9.3, V.10.3).

I.9 Ergodic distributions

Consider a homogeneous process X_k in discrete time modified by a single retaining boundary at $x = 0$ or by replacement when negative values are reached and thereby confined to non-negative states. The process is ergodic under simple conditions (VI.1). The basic requirement is Lindley's condition which says that the first moment $\mu = \int x\, dA(x)$ of the governing increment distribution must be negative. When the process is confined to a finite interval by a second boundary, it will always be ergodic.

For processes on the non-negative half-line $[0,\infty)$, the ergodic distributions are easily exhibited in the real domain when either the negative or positive increments (or both) have exponential distribution (VI.2). One need only make use of the simple properties of the ergodic compensation function $C_\infty(x)$ and ergodic Green measure $G(x)$ developed in Chapter V, for the desired result. The procedure has the advantage of immediate extension to the effects of a second retaining boundary at $x = L$, confining the process to the interval $[0, L]$.

Suppose, for example, the negative increments have exponential distribution, and $A(x)$ is absolutely continuous, so that the extended renewal density $h(x)$ of (7.3) is available. Then the ergodic density $f_\infty(x)$ of the process (2.2) with a single retaining boundary at $x = 0$ may be written in the form

$$f_\infty(x) = \eta|\mu|\{\delta(x) + h(x) - \int_{-\infty}^{0} \eta\, e^{\eta x'} h(x - x')\, dx'\} . \qquad (9.1)$$

When a second retaining barrier is present at $x = L$, it may be seen on simple grounds (VI.3) that the ergodic distribution is given by

$$F_{L\infty}(x) = a_L F_\infty(x), \quad x < L; \qquad F_{L\infty}(x) = 1, \quad L \leqslant x, \tag{9.2}$$

where $F_\infty(x)$ is the distribution for $L = \infty$ corresponding to (9.1). The constant a_L may be determined from a simple self-consistency requirement.

When positive increments have an exponential density $\lambda e^{-\lambda x}$, and there is one retaining boundary at zero, the ergodic distribution is seen to be

$$F_\infty(x) = [1 - \{1 + V_1\lambda^{-1}\} e^{V_1 x}] U(x), \tag{9.3}$$

where V_1 is the negative zero of $1 - \alpha(iV)$. The ergodic distribution for $X_k = \max[0, X_{k-1} + \xi_k]$ when $A(x)$ is more complex may also be obtained (Ch. VI) with the aid of compensation and the ergodic Green measure, but the calculation required is more extensive. Simple asymptotic results will always be available, however. Under simple conditions on $A(x)$ stated in Theorem V.10.1 and Theorem VI.6.1, one finds the asymptotic representation as $x \to \infty$

$$f_\infty(x) \sim \frac{K}{\mu_{V_1}} e^{V_1 x}, \qquad 0 < K < 1. \tag{9.4}$$

Consider next the homogeneous drift process $X_V(t) = vt + \sum_1^{K(t)} \xi_i$ of (4.5). Let us suppose that $v \neq 0$, that the increments have only positive support, and that the homogeneous process is modified by replacement at the value T when the state $x = 0$ is reached. It may then be shown (V.3, VI.5.1) that the ergodic distribution is absolutely continuous with density

$$f_T(x,\infty) = T^{-1} |\nu\mu + v| \{g_c(x-T) - g_c(x)\}, \tag{9.5}$$

where $g_c(x)$ is the ergodic Green density (7.7). If we now consider a sequence of such ergodic processes with T becoming vanishingly small, we find that the limiting process corresponds to a retaining boundary at $x = 0$, and that the corresponding ergodic distribution is

$$F(x,\infty) = \{1 - |\nu\mu + v|\, g_c(x)\} U(x). \tag{9.6}$$

When a second retaining barrier is present at $x = L$, it is quickly inferred from simple structural considerations that

$$F(x,\infty) = \frac{1 - |\nu\mu + v|\, g_c(x)}{1 - |\nu\mu + v|\, g_c(L)} U(x), \qquad x \leqslant L$$

$$F(x,\infty) = 1, \qquad x > L. \tag{9.7}$$

The ergodic distribution for the reservoir of Section 1 is thereby exhibited. When $v > 0$, and increments are negative, a single retaining barrier at $x = 0$ gives rise to an exponential distribution (VI.5).

$$f_\infty(x) = |V_1| e^{V_1 x} U(x) \tag{9.8}$$

where V_1 has the customary meaning of the value in the convergence interval at which $\alpha(iV_1) = 1$.

When a Wiener-Lévy component is present as in (4.7), or $A(x)$ has both positive and negative support, the ergodic distribution is more complex, but asymptotic information is available.

Similar results may be exhibited for the lattice (VI.4). In particular for a skip-free negative process on the finite lattice interval $[0, L]$, with retaining boundaries at $n = 0$ and $n = L$, and $\mu \neq 0$, one obtains the ergodic state probabilities (VI.4, VII.7)

$$p_n^{(\infty)} = \frac{g_n - g_{n+1} + |\mu_{\rho_1}|^{-1} \rho_1^{-n-1}(1-\rho_1)}{|\mu_{\rho_1}|^{-1} \rho_1^{-L-1} - g_{L+1}} . \tag{9.9}$$

When $\mu > 0$, ρ_1 is the real positive zero of $1 - \epsilon(u)$ inside the unit circle, and μ_{ρ_1} is the mean increment for the conjugate increment distribution associated with $\rho = \rho_1$. When $\mu < 0$, $\rho_1 = 1$, $\mu_{\rho_1} = \mu$ and (9.9) reduces to $p_n^{(\infty)} = (g_n - g_{n+1})/(g_0 - g_{L+1})$. For the cumulative probability, $P_n^{(\infty)} = \sum_0^n p_j^{(\infty)}$, we find for the latter case,

$$P_n^{(\infty)} = \frac{1 - |\mu| g_n}{1 - |\mu| g_{L+1}} , \qquad 0 \leqslant n \leqslant L . \tag{9.10}$$

The form of (9.10) may be compared with that of (9.7), and their structural similarity noted. This similarity may be understood when one recognizes that the drift process is a limiting form of the lattice process in continuous time.

I.10 The gambler's ruin problem and the distribution of extremes

Consider the following gambler's ruin problem on the finite interval. A homogeneous process $X(t)$ in discrete or continuous time commences at some point y of the finite half-open interval $[0,L)$. Process samples which reach $(-\infty,0)$ before $[L,\infty)$ are regarded as successes. Samples which reach $[L,\infty)$ before $(-\infty,0)$ are regarded as failures. One seeks the probability $R_y(t)$ of failure at or before time t, and the ruin probability R_y of ultimate failure.

It is somewhat surprising that these probabilities, associated with a passage process, may be obtained from the distribution of the ergodic process resulting when the homogeneous process is modified by retaining boundaries at $x = 0$ and $x = L$ and reversal of its increments.

The ruin probability in particular is available directly from the ergodic distribution. The basic result for processes in discrete time on the continuum is the following:

Theorem VI.7.2. For the homogeneous process $X_k = y + \sum_1^k \xi_i$, with increment distribution $A(x) = P\{\xi_i \leqslant x\}$, the probability $R_y^{(k)}$ of ruin at or before the kth step is identical with the distribution $F_k'(x)$ for X_k', the process with adjoint increment distribution $\widetilde{A}(x) = 1 - A(-x)$, modified by retaining barriers at $x = 0$ and $x = L$, and commencing at $X_0' = L$; i.e.,

$$R_y^{(k)} = F_k'(y). \tag{10.1}$$

The probability of ultimate ruin is the ergodic distribution

$$R_y = F_\infty'(y). \tag{10.2}$$

The basic structure of the result carries over for any homogeneous process in continuous time by virtue of DeFinetti's Theorem. Results for lattice processes and processes in continuous time are given in Theorems VI.7.1 and VI.7.3.

One may also consider a ruin problem on the half-line $[0,\infty)$ for which escape to infinity constitutes ruin. The probability that a process X_k with $\mu > 0$ commencing at $y > 0$ will ever reach $(-\infty,0)$ is obtained from (10.2), and given by $1 - R_y = 1 - F_\infty'(y)$ where $F_\infty'(y)$ is the ergodic distribution for the adjoint process modified by a single retaining boundary at zero.

These ruin results lead immediately to the distribution of extremes for bounded and unbounded processes. Thus for a homogeneous process X_k with mean increment $\mu < 0$ and initial value $x_0 = 0$, the largest value attained subsequently (VI.8) has the distribution

$$E(x) = F_\infty^*(x), \tag{10.3}$$

where $F_\infty^*(x)$ is the ergodic distribution for the process X_k modified by a retaining boundary at $x = 0$.

Of equal or greater practical importance is the distribution of the maximum value ζ_y for a homogeneous process X_k commencing at $y > 0$ attained before reaching the set $(-\infty,0)$. Such random variables as the maximum value between depletions of a bounded process (Example 5 of Section 1), may be discussed thereby.

Theorem VI.8.2. Let X_k be the homogeneous process $X_k = y + \sum_1^k \xi_i$, governed by the increment distribution $A(x)$. Let ζ_y be the maximum value attained by X_k before reaching the open set $(-\infty,0)$, and let

$E_y(z) = P\{\zeta_y \leqslant z\}$. Then

$$E_y(z) = F_\infty^*(z-y), \qquad 0 < y, \tag{10.4}$$

where $F_\infty^*(x)$ is the ergodic distribution for the homogeneous process X_k when modified by retaining boundaries at $x = 0$ and $x = z$.

The ergodic Green distributions exhibited in the previous section for a process on a finite interval may then be employed. The result for a process on the lattice with unit increments in the negative direction and arbitrary increments in the positive direction has a simple form when $\mu < 0$. One finds

$$P\{\zeta_m \leqslant N\} = \frac{g_0 - g_{N+1-m}}{g_0 - g_{N+1}}\, U(N-m), \tag{10.5}$$

where g_n is the ergodic Green mass for the underlying homogeneous process.

I.11 Processes with representation of finite dimension

The final chapter, VII, is concerned with the treatment of processes on the half-line $[0,\infty)$ and finite interval $[0,L]$ when the increment distribution $A(x)$ governing the underlying homogeneous process has a certain kind of structure.

For processes on the half-line the structure is identical with the class of distributions permitting finite factorization of the kernels for the associated Wiener-Hopf and Hilbert problems (V.12). The basic requirement is that increments of one sign have distribution of general exponential form, e.g., have an Erlang distribution, or a mixture of a finite number of such distributions. When negative increments have such distribution one requires more specifically that $A(x)$ be of the form, for $x < 0$,

$$A(x) = \int_{-\infty}^{x} \sum_{1}^{K} e^{\eta_j x'} [P_{1j}(x') \cos \mu_j x' + P_{2j}(x') \sin \mu_j x']\, dx', \tag{11.1}$$

where $\mu_1 = 0$ and $P_{1j}(x)$ and $P_{2j}(x)$ are polynomials in x. One then finds (VII.1) that the local density $a^-(x) = A'(x)\{1-U(x)\}$ has a finite decomposition $a^-(x-x') = \sum_1^{\Delta} a_m(x)\, b_m(x')$ when $x' > 0$ and $x < 0$, for which the functions $a_m(x)$ are linearly independent. This set of Δ functions $a_m(x)$ provides a finite basis of dimension Δ for the discussion of the process, which is an analogue of the Wiener-Hopf factorization.

For the corresponding process on the non-negative half lattice (VII.6) a finite basis is available when the negative increments have a general geometric form, i.e., when, for $n < 0$,

$$e_n = \sum_{k=1}^{K} \beta_k^{|n|}[P_{1k}(n) \cos \mu_k n + P_{2k}(n) \sin \mu_k n], \qquad (11.2)$$

where $\mu_1 = 0$, $0 < \beta_k < 1$, and $P_{1k}(n)$ and $P_{2k}(n)$ are polynomials in n. A finite representation is also available (VII.6) if the negative increments are bounded, i.e., if $e_n = 0$ for $n < -N_1$. The latter case, of strong practical interest, may be regarded as a limiting form of (11.2).

The vector and matrix equations associated with these finite representations are real and, for lattice processes in particular, are well suited to modern computation techniques when solutions in closed form are unavailable.

For processes on $[0,\infty)$ for which the positive increments have a general exponential or geometric form, and for processes on a finite interval, a modification of the compensation technique employed in the first six chapters is required (VII.5, VII.6, VII.7). The following representative result is obtained for the finite lattice:

Theorem VII.7.1. For any replacement process on the finite lattice $\{n: 0 \leqslant n \leqslant L\}$ whose governing increment distribution has one or both bounds (N_1, N_2) finite, a finite representation is available for the discussion of the process of dimension $\Delta = 2N$ where $N = \min\{N_1, N_2\}$.

In particular, for a skip-free process on the finite lattice, $N = 1$ and $\Delta = 2$. The ergodic distribution of (9.9) is obtained in this manner.

CHAPTER II

HOMOGENEOUS MARKOV PROCESSES

This chapter is devoted to development of the elementary structural properties of Markov processes and their governing equations of continuity required for subsequent chapters. Of particular interest are the spatially homogeneous processes on the continuum and lattice, and the duality between Markov processes in discrete time and the associated jump processes in continuous time. The homogeneous diffusion process and drift processes are exhibited as limiting forms of the jump processes. Some properties of the transition functions (Green's functions) for the homogeneous processes are reviewed.

II.1 Stochastic processes: some definitions

A familiarity with the elements of probability theory and of the concept and basic properties of a stochastic process† will be assumed of the reader. For the purposes of this book, a stochastic process $X(t)$ is an ensemble $\{x_\omega(t)\}$ of process sample functions. Each sample is identified by its index ω, and assumes a succession of values x, chosen from some state space X, successive values being indexed by the time parameter t. The distribution of the samples over X at an initial time, and the probabilistic laws governing changes in the sample functions in this state space with time, characterize the process.

---

† The reader may wish to call upon the following sources for background and supplementary material:

Bartlett, M.S. (1955), *An Introduction to Stochastic Processes*, Cambridge University Press.

Bharucha-Reid, A.T. (1960), *Elements of the Theory of Markov Processes and their Applications*, McGraw-Hill, New York.

Doob, J.L. (1953), *Stochastic Processes*, John Wiley & Sons, New York.

Feller, W. (1957), *An Introduction to Probability Theory and its Applications*, John Wiley & Sons, New York.

Fisz, M. (1963), *Probability Theory and Mathematical Statistics*, John Wiley & Sons, New York.

Parzen, E. (1962), *Stochastic Processes*, Holden-Day, Inc., San Francisco, California.

Rosenblatt, M. (1962), *Random Processes*, Oxford University Press.

Proper discussion of a stochastic process normally requires an exposition of the structure of the space of sample functions of the process. Because of the simplicity of the processes treated and the character of the probabilistic information sought, such an exposition is unnecessary and will not be given. All information desired is obtained from the sequence of distributions in time describing the process.

In the early chapters of the book, we will be concerned exclusively with Markov processes in one dimension. A process $X(t)$ will have as its state space the set $\mathcal{X}$ of real numbers or some subset thereof. When all real values of x or the values of some finite or semi-infinite interval are assumed, we shall speak of a *continuum process* or process on the continuum. When only integral values are assumed, the process will be called a *lattice process*.

The time parameter t indexing successive times of interest will itself take on either a continuous set of real values or some sequence of discrete values $k = 0,1,2,\ldots$. We will speak correspondingly of *processes in continuous time* and *processes in discrete time*.

Our principal concern will be with *temporally homogeneous processes*. For such processes, transition probabilities are invariant with time. All processes discussed will be temporally homogeneous unless otherwise stated.

II.2 Markov processes in discrete time and the associated jump processes in continuous time

A Markov process X_k, $k = 0,1,2,\ldots$ in discrete time on the continuum is characterized by (a) a prescribed *initial distribution* $F_0(x) = P\{X_0 \leqslant x\}$ and (b) a prescribed *single-step transition distribution*

$$A(x', x) = P\{X_{k+1} \leqslant x | X_k = x'\}. \tag{2.0}$$

The distribution of X_k, denoted by $F_k(x)$, is generated by the initial distribution $F_0(x)$ and the continuity equation†

$$F_{k+1}(x) = \int A(x', x)\, dF_k(x'). \tag{2.1}$$

Let $A^{(k)}(x', x)$ designate the k-fold iteration of $A(x', x)$, i.e.,

---

† Generality requires that Lebesgue-Stieltjes integration be employed. A familiarity with Riemann-Stieltjes integration will enable the reader to follow the discussion. For an account of the Lebesgue-Stieltjes integral, see Loève (1960).

$$A^{(0)}(x', x) = U(x-x'); \quad A^{(k)}(x', x) = \int A^{(k-1)}(y, x)\, d_y A(x', y)$$

$$\text{for } k \geqslant 1, \qquad (2.2)$$

where $U(x)$ is the degenerate distribution: $U(x) = 0$, $x < 0$; $U(x) = 1$, $x \geqslant 0$. From (2.1) and (2.2), we then have for all k

$$F_k(x) = \int G_k(x', x)\, dF_0(x') \qquad (2.3)$$

where

$$G_k(x', x) = A^{(k)}(x', x). \qquad (2.4)$$

The function $G_k(x', x)$ in (2.3) is the *k-step transition distribution* of the process, i.e., $G_k(x', x) = P\{X_k \leqslant x | X_0 = x'\}$.

A Markov *jump process* $X(t)$ in continuous time is characterized† by an initial distribution $F_0(x)$, a transition frequency ν, and a single-step transition distribution $A(x', x)$. If a sample $x_\omega(t)$ is in state x' at time t, then at time $t+dt$ there is probability $1 - \nu dt$ that the sample has not departed from state x', and probability $\nu\, dt$ that a transition has taken place. If there has been a transition, the sample has distribution $A(x', x)$. Thus the jump process is compounded of two independent processes, a Poisson process $K(t)$ with intensity ν giving the number of transitions that have taken place in $(0,t)$ and a discrete time process $X_k = X(t_k+)$ obtained from $X(t)$ by sampling at the instants t_k+ after the transition epochs. The jump process $X(t)$ governed by $[F_0(x), A(x', x), \nu]$ will be said to be *associated with* the discrete time process X_k governed by $[F_0(x), A(x', x)]$. A sample $x_\omega(t)$ of the process $X(t)$ might

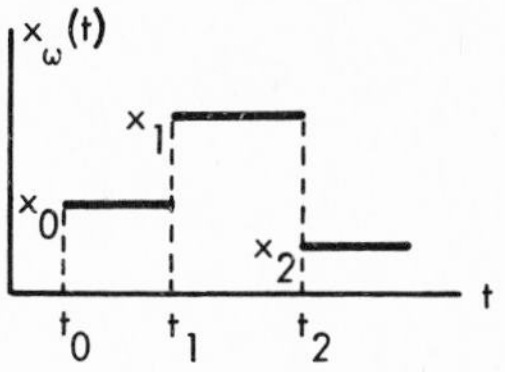

Fig.II.1 A typical process sample for a Markov jump process in continuous time

have the appearance shown in Fig.II.1. Let $\alpha_k(t) = P\{K(t) = k\} = (\nu t)^k e^{-\nu t}/k!$. The jump process $X(t)$ will then have for its distribution, $F(x,t) = P\{X(t) \leqslant x\}$,

$$F(x,t) = \sum_0^\infty \alpha_k(t) F_k(x) = \sum_0^\infty \{e^{-\nu t}(\nu t)^k/k!\} F_k(x) \qquad (2.5)$$

† See Note 1 at the end of the book.

where $F_k(x)$ is given by (2.3). Since $F_k(x) \leqslant 1$ and $\sum \alpha_k(t) = 1$, the convergence of (2.5) is clear. The convergence of the power series for all complex values of t implies that $F(x,t)$ is an entire function of t. From (2.3), (2.4) and (2.5), we have

$$F(x,t) = \int G(x', x, t)\, dF_0(x'), \tag{2.6}$$

where $G(x', x, t)$ is the transition distribution

$$G(x', x, t) = \sum_0^\infty \alpha_k(t)\, G_k(x', x) = \sum_0^\infty \{(\nu t)^k e^{-\nu t}/k!\}\, A^{(k)}(x', x). \tag{2.7}$$

Since there is probability $\nu\, dt$ of a transition in the interval $(t, t+dt)$, a simple continuity argument gives for $k = 0,1,2,\ldots$

$$\frac{d}{dt}\alpha_k(t) = -\nu\, \alpha_k(t) + (1-\delta_{k,0})\, \nu\, \alpha_{k-1}(t). \tag{2.8}$$

If we differentiate (2.5) with respect to time, and employ (2.1) and (2.8), we find that $F(x,t)$ obeys the continuity equation

$$\frac{\partial F(x,t)}{\partial t} = -\nu F(x,t) + \nu \int A(x', x)\, dF(x', t). \tag{2.9}$$

The term-by-term differentiation of (2.5) is justified by the entire character of the power series. Equation (2.9) may be obtained directly from a simple continuity argument, as we will see in Section 3.

The iteration structure of (2.2) implies that, for all values $j = 0,1,\ldots k$, $A^{(k)} = A^{(j)} . A^{(k-j)}$, where the dot denotes the spatial operation of (2.2). One then has the *Chapman-Kolmogorov equation*

$$G_k(x', x) = \int G_{k-j}(x'', x)\, dG_j(x', x''), \qquad 0 \leqslant j \leqslant k. \tag{2.10}$$

The Chapman-Kolmogorov equation for the associated processes in continuous time is

$$G(x', x, t) = \int G(x'', x, t-s)\, dG(x', x'', s), \qquad 0 \leqslant s \leqslant t. \tag{2.11}$$

Equations (2.10) and (2.11) express the temporally homogeneous Markov character of their respective processes. Equation (2.11) for example states that if a sample is at x' at $t = 0$ and at x''' at time t, then at any intervening time s, it must have been in some state x'' and have gone from there to x''' in time $t-s$. It is of interest to note that (2.11) follows from (2.10) (i.e., from $A^{(k)} = A^{(j)} \cdot A^{(k-j)}$), (2.7), and from the identity

$$\frac{t^k}{k!} = \sum_{j=0}^{k} \frac{(t-s)^j}{j!} \frac{s^{k-j}}{(k-j)!},$$

obtained from the binomial expansion of $\{(t-s)+s\}^k = t^k$.

A more general Markov process in continuous time is characterized by a transition frequency ν, a single-step transition distribution $A(x', x)$ and some equation of motion

$$\frac{dx}{dt} = \psi(x) \tag{2.12}$$

governing all process samples between transition epochs.

For all of the processes described above, the transition epochs are isolated and denumerable. A large class of Markov processes in continuous time cannot be characterized in terms of isolated transitions. The diffusion processes are contained in this class. Such processes, however, may be shown to be limiting forms of the jump processes when increments decrease in size and increase in frequency in a certain way. A discussion of such general homogeneous processes will be given in Sections 3 and 7.

A more extensive treatment of temporally and spatially inhomogeneous Markov processes may be found in Doob (1953), Bharucha-Reid (1960), and Rosenblatt (1962). A survey of temporally homogeneous Markov processes on a denumerable state space (Markov chains) has been given by Chung (1960).

II.2.1 Two classes of Poisson events in competition

Of frequent interest is a jump process $X(t)$ whose transitions are initiated by two mutually exclusive classes of events $\mathfrak{E}_1$ and $\mathfrak{E}_2$ each having Poisson statistics, and each associated with its own single-step transition distribution. It is often advantageous to combine the two classes of events $\mathfrak{E}_1$ and $\mathfrak{E}_2$ into a single class of events $\mathfrak{E}$ of frequency $\nu = \nu_1 + \nu_2$, and to regard each event of $\mathfrak{E}$ as having the probabilities $p_1 = \nu_1/\nu$ and $p_2 = \nu_2/\nu$ of being of type 1 and type 2 respectively. If the single-step transition distribution associated with events $\mathfrak{E}_1$ is $A_1(x', x) = P\{X_{k+1} \leqslant x' | X_k = x'$, event of type $1\}$ and that with events $\mathfrak{E}_2$ is $A_2(x', x)$, then $X(t)$ is a Markov jump process in continuous time governed by frequency

$$\nu = \nu_1 + \nu_2 \tag{2.13}$$

and the single-step transition distribution

$$A(x', x) = (\nu_1/\nu) A_1(x', x) + (\nu_2/\nu) A_2(x', x). \qquad (2.14)$$

An example describing the length $X(t)$ of a bus queue to which passengers arrive in groups of random size at Poisson epochs, and to which buses of random capacity arrive at independent Poisson epochs, will be given in Section V.1.

II.3 Spatially homogeneous processes

(a) *Discrete time processes*

In applications of probability theory, processes whose transitions arise from random increments are of constant interest. The sum of such random increments ξ_k arriving at discrete epochs t_k is the stochastic process

$$X_k = X_0 + \sum_{i=1}^{k} \xi_i, \qquad (3.1)$$

which may be described as being additive. When the random increment ξ_k has a distribution dependent only on the previous partial sum X_{k-1}, the process X_k is temporally homogeneous and Markov. When the increments ξ_k are independent and identically distributed, the single-step transition distribution $A(x', x)$ of Section 2 depends only on $x - x'$ and $A(x', x) = A(0, x-x')$. Let $A(0,x)$, *the increment distribution*, be denoted simply by $A(x)$. Since $A(x', x) = A(x-x')$, the kernel for the iterations (2.2) is of *convolution type*, and $A^{(k)}(x', x)$ is the k-fold convolution $A^{(k)}(x)$ of the increment distribution with itself. $A^{(k)}(x)$ is defined recursively by $A^{(0)}(x) = U(x)$ and

$$A^{(k+1)}(x) = \int_{-\infty}^{\infty} A^{(k)}(x-x')\, dA(x'). \qquad (3.2)$$

The k-step transition distribution $G_k(x', x)$ is, by (2.4), $A^{(k)}(x', x) = A^{(k)}(x-x')$. Let $G_k(0,x)$ be denoted by $G_k(x)$. For the process X_k in discrete time, governed by the increment distribution $A(x)$, (2.3) takes the form

$$F_k(x) = \int_{-\infty}^{\infty} G_k(x-x')\, dF_0(x') = \int_{-\infty}^{\infty} A^{(k)}(x-x')\, dF_0(x'). \qquad (3.3)$$

(b) *Jump processes*

The process $X(t)$ in continuous time, governed by the same increment distribution $A(x)$ and increment frequency ν, may be characterized by the equation

$$X(t) = X_0 + \sum_{1}^{K(t)} \xi_i \tag{3.4}$$

where $K(t)$ is the auxiliary Poisson process (Section 2) of frequency ν. Such a process is often referred to as a *compound Poisson process* (cf. Parzen, 1962). The probability transition distribution of (2.7) now has the form $G(x', x, t) = G(x-x', t)$, where from (2.7)

$$G(x,t) = \sum_{k=0}^{\infty} e^{-\nu t} \frac{(\nu t)^k}{k!} A^{(k)}(x). \tag{3.5}$$

The distribution of $X(t)$, given from (2.6) by

$$F(x,t) = \int_{-\infty}^{\infty} G(x-x', t)\, dF_0(x'), \tag{3.6}$$

satisfies the continuity equation (2.9) which becomes

$$\frac{\partial}{\partial t} F(x,t) = -\nu F(x,t) + \nu \int_{-\infty}^{\infty} A(x-x')\, dF(x', t). \tag{3.7}$$

Equations (3.3) and (3.7) are homogeneous in x, and the additive processes with independent random increments are said to be *spatially homogeneous processes.*

If the mean and variance of the increment distribution $A(x)$ exist and are given by μ_1 and σ_1^2 respectively, then the mean and variance of $G_k(x)$ are

$$\mu_k = k\mu_1 \quad \text{and} \quad \sigma_k^2 = k\sigma_1^2. \tag{3.8}$$

The jump process $X(t)$ of frequency ν and increment distribution $A(x)$ will then also have for its transition distribution $G(x,t)$ a mean and variance, given by

$$\mu(t) = \nu t \mu_1 \quad \text{and} \quad \sigma^2(t) = \nu t(\sigma_1^2 + \mu_1^2). \tag{3.9}$$

We note that the distribution $G(x,t)$ for a jump process of frequency ν and increment distribution $A(x)$ acquires some of the simple properties of $A(x)$. In particular, if $A(x)$ is absolutely continuous, then $G(x,t)$ is absolutely continuous in every interval not containing the origin. For, from (3.5),

$$G(x,t) = U(x)e^{-\nu t} + A(x) \cdot \left[\sum_{1}^{\infty} \{ e^{-\nu t} (\nu t)^k / k! \} A^{(k-1)}(x) \right],$$

and the second term on the right is absolutely continuous.

When discussing probability densities we will employ the notation convention of designating the densities by lower-case letters, e.g., when the distribution $A(x)$ of increments is absolutely continuous, the corresponding probability density will be denoted by $a(x)$, so that $A(x) = \int_{-\infty}^{x} a(x')\,dx'$.

A second less obvious property may now be stated.

Theorem II.3.1. If $A(x)$ is absolutely continuous and its probability density $a(x)$ has finite total variation $V[a]$ on $(-\infty, \infty)$, then the transition density with singular component removed,

$$g(x,t) - \delta(x)\,e^{-\nu t} = \sum_{1}^{\infty} \{e^{-\nu t}(\nu t)^k/k!\}\,a^{(k)}(x),$$

has a finite total variation on $(-\infty, \infty)$ smaller than $V[a]$.

Let a subdivision Δ of the interval $(-\infty, \infty)$ be specified by an infinite, bilateral sequence x_n: $n = \ldots -1, 0, 1, \ldots$ where $x_n < x_{n+1}$ for all n. For any function $c(x)$, let $V_\Delta[c] = \sum_n |c(x_n) - c(x_{n+1})|$ be the variation of $c(x)$ for subdivision Δ. The *total variation* $V[c]$ of $c(x)$ on $(-\infty, \infty)$ is the least upper bound of $V_\Delta[c]$ for subdivisions Δ. We will prove the theorem with the aid of the following lemma.

Lemma II.3.1. Let $a(x)$ and $b(x)$ be two probability densities with total variations $V[a]$ and $V[b]$ on $(-\infty, \infty)$, at least one of which is finite. Let $a \cdot b$ be the convolution $\int_{-\infty}^{\infty} a(x')\,b(x-x')\,dx'$ of $a(x)$ and $b(x)$. Then

$$V[a \cdot b] \leqslant \min\{V[a], V[b]\}. \tag{3.10}$$

To prove this, let $V[a]$ be finite. For any subdivision Δ, one has for $V_\Delta[a \cdot b]$,

$$\sum_n \left| \int \{a(x_n - x') - a(x_{n+1} - x')\}\,b(x')\,dx' \right|$$

$$\leqslant \sum_n \int \left| a(x_n - x') - a(x_{n+1} - x') \right| b(x')\,dx'$$

$$= \int \sum_n \left| a(x_n - x') - a(x_{n+1} - x') \right| b(x')\,dx'$$

$$\leqslant V[a] \int b(x')\,dx' = V[a].$$

The interchange of summation and integration is justified by the absolute convergence of the summation-integration. From $V_\Delta[a \cdot b] \leqslant V[a]$, it follows that $V[a \cdot b] \leqslant V[a]$. Similarly, if $V[b]$ is finite, $V[a \cdot b] \leqslant V[b]$. We then have (3.10).

To prove the theorem we observe that, for any constants p_n and functions $a_n(x)$, we have $V[\Sigma p_n a_n(x)] \leqslant \Sigma |p_n|\, V[a_n]$. Since $V[a^{(k)}(x)] \leqslant \min\{V[a],\ V[a^{(k-1)}(x)]\} \leqslant V[a]$, and $\sum_1^\infty \{e^{-\nu t}(\nu t)^k/k!\} < 1$, the theorem follows.

(e) *Drift processes*

A slightly more general spatially homogeneous Markov process is obtained when a drift term vt, linear in time, is added to the compound Poisson process $X(t)$ of (3.4), giving a new *drift process*

$$X_v(t) = X_0 + vt + \sum_{i=1}^{K(t)} \xi_i = X(t) + vt. \tag{3.11}$$

The sample functions for this process have a constant slope $(d/dt)x_\omega(t) = v$ between transitions, and are discontinuous at the transitions (Fig. II.2).

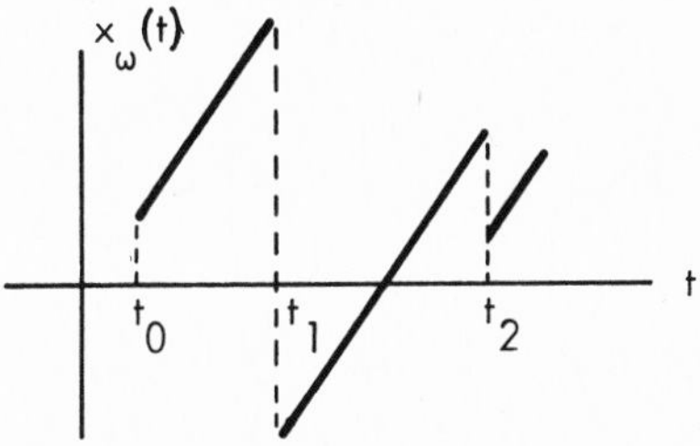

Fig.II.2–A typical sample function for a drift process

A physical realization of such a process might be provided by the amount of water in a reservoir for which the influx due to rainfall may be represented as a compound Poisson process and for which the efflux rate is constant – if the reservoir neither empties nor overflows.† A second example might be obtained from the theory of collective risk for an insurance operation, with $X_v(t)$ representing the capital on hand, vt the relatively smooth input from premiums and ξ_i the sporadic lump sum payments out – when bankruptcy is ignored. The same type of structure appears in connection with the *Takács virtual waiting-time process* of queuing theory, which we will discuss in Chapters IV and V. The distribution $F_v(x,t)$ of $X_v(t)$ is simply related to that for $X(t)$ by $F_v(x,t) = F(x-vt,t)$. The transition distribution is thus obtained from (3.5), i.e., $G_v(x',x,t) = G_v(x-x',t)$ where

† The inhomogeneity in the process due to the boundaries associated with emptying and overflow is of basic interest. The bounded processes will be discussed subsequently.

$$G_V(x,t) = G(x-vt,t) = \sum_0^\infty \{(\nu t)^k e^{-\nu t}/k!\} A^{(k)}(x-vt). \quad (3.12)$$

The equation of continuity satisfied by $F_V(x,t)$ is

$$\frac{\partial}{\partial t} F_V(x,t) + v\frac{\partial}{\partial x} F_V(x,t) = -\nu F_V(x,t) + \nu \int_{-\infty}^{\infty} A(x-x')\, dF_V(x',t). \quad (3.13)$$

Whereas equations (2.9) and (3.7) have a general validity, the spatial derivative present in (3.13) requires that $F_V(x,t)$ be differentiable and this must be supported by some such condition as the differentiability of $F_0(x)$ and $A(x)$. As we will see in equation (3.16), equation (2.9) has a direct analogue for the drift process with the generality of (3.7).

It is often advantageous to work with probability density functions rather than distribution functions. If $F_V(x,t)$ has a density $f_V(x,t)$ which is differentiable in both x and t, spatial differentiation of (3.13) gives

$$\frac{\partial f_V}{\partial t} + v\frac{\partial f_V}{\partial x} = -\nu f_V(x,t) + \nu \int_{-\infty}^{\infty} f_V(x',t)\, a(x-x')\, dx'. \quad (3.14)$$

Even when the governing distributions are not absolutely continuous, equation (3.14) may be meaningful in a generalized sense, such as that of the theory of distributions of L. Schwartz (1950). Because densities have a local character, an equation governing densities will often carry a direct probabilistic interpretation which may facilitate and motivate subsequent analysis. Thus when $v = 0$, equation (3.14) says that the rate of change of the density $f_V(x,t)$ in a small interval of length dx about x is compounded of the rate of loss $\nu f_V(x,t)$ due to transitions from the interval, and the rate of gain due to transitions to the interval from other states x'. When $v \neq 0$, a similar simple interpretation is available. Instead of observing the rate of change at some fixed x, we may have an observer moving with velocity v monitoring the density $f_V(x_0 + vt, t)$ at the position $y_t = x_0 + vt$. The rate of change of density at the moving position of the observer is $\frac{d}{dt} f_V(y_t,t)$, and by the above reasoning this must equal $-\nu f_V(y_t,t) + \nu \int f_V(x',t) \times a(y_t - x')\, dx'$. We have, however,

$$\frac{d}{dt} f_V(x_0 + vt,t) = \left[v\frac{\partial}{\partial x} f_V(x,t) + \frac{\partial}{\partial t} f_V(x,t) \right]_{x = x_0 + vt} \quad (3.15)$$

Replacing $x_0 + vt$ by x, we then have (3.14). The character of the operator $\frac{D}{Dt} = \left(\frac{\partial}{\partial t} + \frac{\partial}{\partial x}\right)$ as a moving differentiation in time is familiar in hydrodynamics. With such differentiation, (3.13) takes the form

$$\frac{D}{Dt} F_v(x,t) = -\nu F_v(x,t) + \nu \int_{-\infty}^{\infty} A(x-x')\, dF_v(x',t) \qquad (3.16)$$

and this equation has a similar *moving observer* interpretation. A solution of (3.16) is (3.12).

(d) *General homogeneous processes; infinite divisibility*

For the homogeneous processes (3.1) and (3.4) in discrete and continuous time, the Chapman-Kolmogorov equations (2.10) and (2.11) take the simpler form

$$G_k(x) = \int G_{k-j}(x-x'')\, dG_j(x''), \qquad 0 \leqslant j \leqslant k; \qquad (3.17)$$

and

$$G(x,t) = \int G(x-x'', t-s)\, dG(x'', s), \qquad 0 \leqslant s \leqslant t. \qquad (3.17')$$

These equations are obtained by setting $x' = 0$ and using $G_k(y,x) = G_k(x-y)$ and $G(y,x,t) = G(x-y,t)$. For the drift process (3.11), it is easy to verify that the transition distribution $G_v(x,t)$ satisfies (3.17′) as a simple consequence of $G_v(x,t) = G_0(x-vt,t)$, and the validity of (3.17′) for $G_0(x,t)$.

The Chapman-Kolmogorov equation (3.17′) may also be obtained from the infinite divisibility of the transition distribution $G(x,t)$ for any homogeneous process in continuous time. A probability distribution $F(x)$ is said to be *infinitely divisible* if for any $n > 1$, $F(x)$ is the n-fold convolution of some other probability distribution $F_n(x)$ with itself. The infinite divisibility of $G(x,t)$ is obvious since $G(x,t)$ is the distribution of the random variable $X(t) - X(0)$, and this may be written as the sum $\sum_0^N [X(t_{n+1}) - X(t_n)]$, where $t_0 = 0$ and $t_{N+1} = t$, for any subdivision $t_0, t_1, \ldots t_{N+1}$ of the interval $(0,t)$. The infinite divisibility is then a consequence of the homogeneity which makes all of the random variables $[X(t_{j+1}) - X(t_j)]$ *independent*. A homogeneous process is often called *a process with independent increments*. An excellent summary of infinite divisibility is given by Lukacs.

The property of infinite divisibility characterizes the most general homogeneous Markov process in continuous time. For every infinitely divisible distribution $G(x)$, there will be a homogeneous process with transition distribution $G(x,t) = G^{(t)}(x)$, the t-fold convolution of $G(x)$ with itself. Such a convolution with t a continuous parameter is made meaningful for infinitely divisible distributions via the characteristic function of $G(x)$ (Section 7).

The simplest and most important example is provided by the Gaussian distribution with density $g(x) = (4\pi D)^{-\frac{1}{2}} \exp\{-x^2/4D\}$ for which

$$g_D^{(t)}(x) = (4\pi Dt)^{-\frac{1}{2}} \exp\{-x^2/4Dt\}. \tag{3.18}$$

A process of independent increments with $X_0 = 0$ and $g(x,t) = g_D^{(t)}(x)$ is called a *Wiener-Lévy process.* A simple indication that sample functions of the Wiener-Lévy process are continuous with probability one will be given in Section 7. The Wiener-Lévy process is the only homogeneous process with this property, terms $X_0 + vt$ apart.

Another example of many that may be derived from the infinitely divisible distributions is that associated with the exponential distribution $G(x) = (1 - e^{-\lambda x})U(x)$, whose t-fold convolution is a Gamma distribution with the density

$$g(x,t) = g^{(t)}(x) = \frac{\lambda(\lambda x)^{t-1} e^{-\lambda x} U(x)}{\Gamma(t)}. \tag{3.19}$$

The homogeneous process with transition density (3.19) is of interest in its own right since it describes a simple type of Brownian motion with positive increments.

It will be convenient subsequently to treat a combined process

$$X(t) = X_0 + X_J(t) + X_D(t) + vt, \tag{3.20}$$

where X_0 is the initial value, $X_J(t)$ is a jump process (3.4) with $X_J(0) = 0$, and $X_D(t)$ is a Wiener-Lévy process. The corresponding transition distribution of $X(t)$ will be the convolution of $F_0(x)$ and the transition distributions for the other components. When the Wiener-Lévy component is present, the distribution of $X(t)$ will inherit the absolute continuity for $t > 0$ of that for $X_D(t)$. The density $f(x,t)$ of $X(t)$ will be shown in Section 7 to satisfy the equation of continuity

$$\left(\frac{\partial}{\partial t} + v\frac{\partial}{\partial x}\right) f(x,t) = D\frac{\partial^2 f(x,t)}{\partial x^2} - \nu f(x,t) + \nu \int f(x-x', t)\, dA(x'), \tag{3.21}$$

where ν and $A(x)$ are associated with the jump process $X_J(t)$. When the component $X_J(t)$ is zero, the process $X(t)$ will be called a *homogeneous diffusion process*. Equation (3.21) then reduces to the homogeneous diffusion equation

$$\frac{\partial}{\partial t} f(x,t) = D \frac{\partial^2 f(x,t)}{\partial x^2} - v \frac{\partial}{\partial x} f(x,t). \tag{3.22}$$

The transition density $g_D(x,t)$ of (3.14) is a solution of this equation for $v = 0$ and the initial density $f(x,0) = \delta(x)$. Correspondingly, for the same initial density, (3.22) will be satisfied by the transition density $g(x,t) = g_D(x,t)$, i.e., by

$$g(x,t) = (4\pi D t)^{-\frac{1}{2}} \exp\{ -(x-vt)^2/4Dt\}. \tag{3.23}$$

It should be noted that not every homogeneous process will satisfy an integro-differential equation. The transition density (3.19) for example does not satisfy such an equation.

The spatially homogeneous processes have central importance in subsequent chapters. As we shall see in Chapter V, the transition functions $G_k(x)$ and $G(x,t)$ for these spatially homogeneous processes play a key role in the treatment of associated bounded processes. The role played is that of source functions or *Green's functions* for the effective sources introduced by the inhomogeneities associated with the boundaries. To call attention to this role we will often refer to the transition function for the homogeneous process as the Green's function for the bounded process. For clarity, we will sometimes speak of the Green distribution $G(x,t)$ or Green density $g(x,t)$ where appropriate. For homogeneous lattice processes, discussed in Section 4, we will speak of *Green probabilities*.

II.4 Lattice processes

When the initial distribution $F_0(x)$ has its support on the lattice of integers and transitions terminate in integral values, the discrete time process X_k of Section 2 is confined to the lattice. It will then be denoted by N_k. Markov processes in discrete or continuous time on a denumerable state space are called *Markov chains*.

The distribution of N_k may be written in the lattice form

$$F_k(x) = \sum_{-\infty}^{\infty} p_n^{(k)} U(x-n), \tag{4.1}$$

where $p_n^{(k)}$ is the state probability $p_n^{(k)} = P\{N_k = n\}$. (Because these state probabilities have a local character in contrast to the distribution which has a cumulative character, they will be designated by lower-case fount.) Correspondingly, the single-step transition distribution $A(n', x)$ may be given the form $A(n', x) = \sum_{-\infty}^{+\infty} \mathrm{e}_{n'n} U(x-n)$, where $\mathrm{e}_{n'n}$ is the *single-step transition probability* $\mathrm{e}_{n'n} = P\{N_{k+1}=n | N_k = n'\}$. Since $A(n', x)$ is a distribution, one has $\sum_n \mathrm{e}_{n'n} = 1$ for all n'. The k-step transition probabilities $g_{n'n}^{(k)}$ are the iterations of $\mathrm{e}_{n'n}$ defined as in (2.2), i.e.,

$$g_{n'n}^{(k)} = \mathrm{e}_{n'n}^{(k)}$$

where

$$\mathrm{e}_{n'n}^{(0)} = \delta_{n'n}; \quad \mathrm{e}_{n'n}^{(k)} = \sum_{n''} \mathrm{e}_{n'n''}^{(k-1)} \mathrm{e}_{n''n} \qquad \text{for } k \geqslant 1. \tag{4.2}$$

In this notation the recursion relation (2.1) becomes

$$p_n^{(k+1)} = \sum_{n'} p_{n'}^{(k)} g_{n'n}^{(1)} = \sum_{n'} p_{n'}^{(0)} g_{n'n}^{(k+1)}. \tag{4.3}$$

Associated with the process N_k in discrete time is the jump process $N(t)$ in continuous time governed by the same single-step transition probabilities $\mathrm{e}_{n'n}$ and transition frequency ν. The distribution of $N(t)$ becomes

$$F(x,t) = \sum_{-\infty}^{\infty} p_n(t)\, U(x-n) \tag{4.4}$$

with state probabilities $p_n(t) = P\{N(t) = n\}$. The continuity equation (2.9) then takes the form

$$\frac{d}{dt} p_n(t) = -\nu p_n(t) + \nu \sum_{n'} p_{n'}(t) \mathrm{e}_{n'n}. \tag{4.5}$$

Equation (4.5) has a simple probabilistic interpretation. The rate of change $dp_n(t)/dt$ of any state probability is the sum of the loss rate $-\nu p_n(t)$ for transitions from the state and gain rate $\nu \sum p_{n'}(t) \mathrm{e}_{n'n}$ for transitions to the state. When diagonal elements e_{nn} are not zero, transitions from states to themselves must be included. The analogue of (2.6) is

$$p_n(t) = \sum_{n'} p_{n'}(0) g_{n'n}(t), \tag{4.6}$$

where $g_{n'n}(t)$, the probability transition function, is

$$g_{n'n}(t) = \sum_{k=0}^{\infty} \left\{ e^{-\nu t} \frac{(\nu t)^k}{k!} \right\} e_{n'n}^{(k)}. \tag{4.7}$$

The Markov lattice process N_k in discrete time may also be described by the analogue of (3.1):

$$N_k = N_0 + \sum_{i=1}^{k} \xi_i , \tag{4.8}$$

where the kth increment ξ_k is a lattice random variable. The Markov character assumed for N_k requires that ξ_k have a distribution dependent at most on N_{k-1}. If the increments are independent with distribution $A(x) = \sum_{-\infty}^{\infty} e_n U(x-n)$, so that $e_n = P\{\xi_k = n\}$, then the single-step transition probabilities are given by

$$e_{n'n} = e_{n-n'} . \tag{4.9}$$

Because $e_{n'n}$ depends only on the difference between n and n', the iterations of (4.2) have a convolution character and give rise, for the k-step transition probabilities $g_{n'n}^{(k)}$, to the structure

$$g_{n'n}^{(k)} = e_{n-n'}^{(k)} , \tag{4.10}$$

where $e_n^{(k)}$ is defined recursively by $e_n^{(0)} = \delta_{n0}$ and

$$e_n^{(k+1)} = \sum_{-\infty}^{\infty} e_j^{(k)} e_{n-j} . \tag{4.11}$$

For the process $N(t)$ in continuous time governed by the same single-step transition probabilities and transition frequency ν, the probability transition function $g_{n'n}(t)$ given by (4.7) may be seen from this equation to have the difference structure $g_{n'n}(t) = g_{n-n'}(t)$, with

$$g_n(t) = \sum_{k=0}^{\infty} \left\{ e^{-\nu t} \frac{(\nu t)^k}{k!} \right\} e_n^{(k)} . \tag{4.12}$$

Equation (4.6) then takes the form

$$p_n(t) = \sum_{-\infty}^{\infty} p_{n'}(0)\, g_{n-n'}(t). \tag{4.13}$$

From equation (4.5), the functions $g_n(t)$ are a solution of the equations

$$\frac{d}{dt} g_n(t) = -\nu g_n(t) + \nu \sum_{n'} g_{n'}(t)\, e_{n-n'} . \tag{4.14}$$

The Chapman-Kolmogorov equations (3.17) and (3.17′) become

$$g_n^{(k)} = \sum_{n'} g_{n-n'}^{(k-j)}\, g_{n'}^{(j)} , \qquad 0 \leqslant j \leqslant k ; \tag{4.15}$$

and

$$g_n(t) = \sum_{n'} g_{n-n'}(t-s)\, g_{n'}(s), \qquad 0 \leqslant s \leqslant t . \tag{4.16}$$

II.5 Reformulation of lattice processes described in terms of transition frequencies

The lattice processes in continuous time are sometimes characterized by a set of transition frequencies $\{\nu_{n'n}\}$. If one knows that a process sample in state n' at time t has probability $\nu_{n'n}dt$ of going to state n in the interval $(t, t{+}dt)$, the continuity argument leading to (4.5) gives the continuity equations

$$\frac{d}{dt} p_n(t) = -\left(\sum_{n'} \nu_{nn'}\right) p_n(t) + \sum_{n'} p_{n'}(t)\nu_{n'n} . \tag{5.1}$$

In such a description, the elements ν_{nn} are taken to be zero. When the rates of transition from state n, $\nu_n = \sum_m \nu_{nm}$, are bounded (as will always be true if there are only a finite number of states), the description in terms of transition frequencies $\nu_{n'n}$ may be reformulated to one in terms of single-step transition probabilities $e_{n'n}$ and an over-all transition frequency ν.† For we may choose ν to be any number $\geqslant \max\{\nu_n\}$ and rewrite (5.1) as

$$\frac{d}{dt} p_n(t) = -\nu p_n(t) + \nu\left[\sum p_{n'}(t)\frac{\nu_{n'n}}{\nu} + p_n(t)\left(\frac{\nu-\nu_n}{\nu}\right)\right]. \tag{5.2}$$

By comparison with (4.5) we see that the original process is equivalent to the process in continuous time governed by the single-step

† A similar discussion in a more general setting may be found in Keilson and Wishart (1964a).

transition probabilities

$$e_{n'n} = \nu^{-1}\nu_{n'n}, \; n' \neq n; \quad e_{nn} = 1 - \nu^{-1}\nu_n; \tag{5.3}$$

and the transition frequency ν.

An example is provided by the process $N(t)$ describing the number of customers in a system when there is probability $\lambda\, dt$ of the arrival of a new customer to the system in any interval of length dt and probability $\eta\, dt$ of a customer departure when the system is not empty. The states consist of the set of non-negative integers $n = 0,1,2,\ldots$,etc. The state probabilities $p_n(t) = P\{N(t)=n\}$ are governed by (5.1) with $\nu_{n,n+1} = \lambda$ for $n \geqslant 0$, $\nu_{n,n-1} = \eta$ for $n \geqslant 1$, and all other transition frequencies zero. Then clearly $\nu_0 = \lambda$, and $\nu_n = \lambda + \eta$ for $n > 1$, so that $\max\{\nu_n\} = \lambda + \eta$, and we may choose this value for ν. We then have from (5.3) the associated single-step transition probabilities

$$e_{on} = \eta\nu^{-1}\delta_{no} + \lambda\nu^{-1}\delta_{n1}. \tag{5.4}$$

and for $n' > 0$,

$$e_{n'n} = \eta\nu^{-1}\delta_{n,n'-1} + \lambda\nu^{-1}\delta_{n,n'+1}. \tag{5.5}$$

We note that $\sum_n e_{n'n} = 1$ for all n' as required. A solution to the system of equations (5.1) is then given by (4.7). The queue governed by these simple laws is called the *Poisson queue*.

A *birth-death* process is any temporally homogeneous Markov process on the non-negative half-lattice for which only increments +1 and −1 are permitted. It is clear that for any such process and more generally for any Markov chain in continuous time for which the rate of departure ν_n from the state n is bounded, the above method provides a solution. The method is well suited to machine computation.

II.6 The characteristic function and generating function

For the homogeneous processes discussed in Sections 3 and 4, analysis is facilitated by the introduction of characteristic functions and generating functions. The *characteristic function* of a distribution $F(x)$, is defined by the expectation

$$\phi(z) = E(e^{izx}) = \int_{-\infty}^{\infty} e^{izx}\, dF(x). \tag{6.1}$$

The integral in (6.1) is a *Fourier-Stieltjes* transform of $F(x)$. When $F(x)$ is absolutely continuous, $\phi(z)$ is the ordinary Fourier transform of the probability density $f(x)$ given by

$$\phi(z) = \int_{-\infty}^{\infty} e^{izx} f(x)\, dx . \tag{6.2}$$

Characteristic functions will be denoted by corresponding lower-case Greek letters. A good derivation of the properties of characteristic functions may be found in a companion volume of this series by E. Lukacs. A tabulation of the more elementary properties *on the real z axis* will be convenient.

(a) For any distribution $F(x)$, the defining integral (6.1) will converge for real z, i.e., the characteristic function will always be available on the real z axis.

(b) $\phi(0) = 1$. If $F(x)$ is absolutely continuous, $\phi(-\infty) = \phi(+\infty) = 0$. If $F(x)$ has any discontinuities, $\phi(z)$ will be oscillatory at infinity on the real axis.

(c) $\phi(-z) = \overline{\phi(z)}$. If the distribution $F(x)$ is symmetric about $x = 0$, i.e., if $F(x) + F(-x) = 1$ ($f(x)$ even), then $\phi(z)$ is real.

(d) $|\phi(z)| \leqslant 1$. The characteristic function $\phi(z)$ will take the value one at $z \neq 0$, only if $F(x)$ is a lattice distribution.† $\phi(z)$ will then be a periodic function.

(e) Every characteristic function is continuous (uniformly) for all real z. If the distribution $F(x)$ has a finite rth moment $\mu_r = \int x^r dF(x)$ (and hence all lower-order moments), the characteristic function can be differentiated r times and

$$(d/dz)^r \phi(z) = i^r \int_{-\infty}^{\infty} x^r e^{izx} dF(x) . \tag{6.3}$$

(f) Two distributions are identical if and only if their characteristic functions are identical. The Inversion Theorem states that for any values a and $a + x > a$ at which $F(x)$ is continuous

$$F(a+x) - F(a) = \lim_{T\to\infty} \frac{1}{2\pi} \int_{-T}^{T} \frac{1 - e^{-izx}}{iz} e^{-iza} \phi(z)\, dz . \tag{6.4}$$

If $\phi(z)$ is absolutely integrable over $(-\infty, \infty)$ then $F(x)$ is absolutely continuous. The associated probability density, given by the inverse Fourier transform

---

† Here lattice distribution is being used in its more general sense of a distribution with all of its support on a set $\{\kappa n: n = 0, \pm 1, \pm 2, \text{etc.}\}$ where κ is real.

$$f(x) = F'(x) = \frac{1}{2\pi}\int_{-\infty}^{\infty} e^{-izx}\,\phi(z)\,dz\,, \tag{6.5}$$

is then continuous for all x. Of frequent importance is the following auxiliary result (Titchmarsh, 1948). If $F(x)$ is absolutely continuous and its density $f(x)$ is of bounded variation in the neighbourhood of $x = x'$, then

$$\frac{1}{2}\{f(x'+0) + f(x'-0)\} = \frac{1}{2\pi}\lim_{T\to\infty}\int_{-T}^{T} e^{-ix'z}\,\phi(z)\,dz\,. \tag{6.5'}$$

(g) The convolution of two distributions $F_1(x)$ and $F_2(x)$ defined by $F(x) = \int F_1(x-x')\,dF_2(x')$ is symmetrical in F_1 and F_2 and is itself a distribution. The characteristic function of a convolution is the product of the characteristic functions, i.e., $\phi(z) = \phi_1(z)\,\phi_2(z)$. Conversely, the product of two characteristic functions is a characteristic function, and its distribution is the convolution of the two distributions.

If equations (3.3), (3.5), (3.6) and (3.7) are subjected to Fourier-Stieltjes transformation, we obtain the equations

$$\phi_k(z) = \alpha^k(z)\,\phi_0(z)\,, \tag{6.6}$$

$$\phi(z,t) = \gamma(z,t)\,\phi_0(z) = e^{-\nu t[1-\alpha(z)]}\,\phi_0(z)\,, \tag{6.7}$$

and

$$\frac{\partial}{\partial t}\phi(z,t) = -\nu[1-\alpha(z)]\,\phi(z,t)\,. \tag{6.8}$$

Similarly for the process (3.11), since $F_V(x,t) = F(x-vt,t)$, we have from (6.7)

$$\phi_V(z,t) = e^{ivzt}\,\phi(z,t) = e^{ivzt}\,e^{-\nu t[1-\alpha(z)]}\,\phi_0(z)\,, \tag{6.9}$$

which satisfies

$$\frac{\partial\phi_V(z,t)}{\partial t} = [ivz - \nu\{1-\alpha(z)\}]\,\phi_V(z,t)\,. \tag{6.10}$$

The characteristic function of a lattice distribution, $F(x) = \sum_{-\infty}^{\infty} p_n\,U(x-n)$, has the form

$$\phi(z) = \sum_{-\infty}^{\infty} p_n\,e^{inz} \tag{6.11}$$

which is a Fourier series, periodic on the real z axis. For lattice distributions, it is convenient and customary to introduce the new transform variable $u = e^{iz}$. The characteristic function then appears as the *two-sided generating function*

$$\pi(u) = \sum_{-\infty}^{\infty} p_n u^n . \tag{6.12}$$

The properties of such two-sided generating functions are the direct analogues of those above for characteristic functions. Specifically we have, on the unit circle:

(a) For any set of lattice probabilities $\{p_n\}$, the series $\pi(u) = \sum_{-\infty}^{\infty} p_n u^n$ converges on the unit circle $|u| = 1$, i.e., the generating function $\pi(u)$ always exists there. It need not exist elsewhere.

(b) $\pi(1) = 1$.

(c) $\overline{\pi(u)} = \pi(u^{-1})$.

(d) $|\pi(u)| \leqslant 1$.

(e) $\pi(u)$ is continuous. If there is a finite rth moment, $\mu_r = \sum_{-\infty}^{\infty} n^r p_n$, $\pi(u)$ can be differentiated r times term by term.

(f) Two sets of lattice probabilities are identical, if and only if their generating functions are identical. Since $\pi(e^{iz})$ is a Fourier series, its coefficients are given by the inversion formula

$$p_n = \frac{1}{2\pi}\int_0^{2\pi} \pi(e^{iz})e^{-inz}\,dz = \frac{1}{2\pi i}\oint \pi(u)\frac{du}{u^{n+1}} \tag{6.13}$$

where the contour for the complex integral on the right is the unit circle $|u| = 1$, taken in the positive sense. When $\pi(u)$ has a suitable analytic structure, (6.13) may be recognized as the formula for the coefficients in a Laurent series.

(g) The convolution of two lattice distributions is a lattice distribution with $p_n = \sum_{n'} p_{1\,(n-n')}\, p_{2n'}$. The generating function is the product of the two generating functions, i.e., $\pi(u) = \pi_1(u)\,\pi_2(u)$.

If we denote the generating function of the increment probabilities e_n for the homogeneous lattice processes of Section 4 by $\epsilon(u)$, and that of $g_n(t)$ in equation (4.12) by $\gamma(u,t)$, the generating function $\gamma(u,t)$ is found from (4.12) to be

$$\gamma(u,t) = e^{-\nu t[1-\epsilon(u)]} . \tag{6.14}$$

Correspondingly, for the generating function of the state probabilities, equation (4.13) becomes

$$\pi(u,t) = \pi(u,0)\,\gamma(u,t) = \pi(u,0)e^{-\nu t[1-\epsilon(u)]}. \qquad (6.15)$$

II.7 Diffusion processes and drift processes as limiting forms

The homogeneous Markov jump processes $X(t)$, characterized by (3.4), share the simplicity of the homogeneous processes (3.1) in discrete time, and as we will see, share many of the structural properties associated with this simplicity. An important theorem in the theory of infinitely divisible processes permits one to express any homogeneous process in continuous time as the limit of a sequence of such jump processes. In particular, for drift processes and processes with a Wiener-Lévy component, a simple representation of this form is available, permitting a substantial economy and uniformity of exposition in later chapters.

If $G(x)$ is an infinitely divisible distribution with characteristic function $\gamma(z)$, the function $\gamma^t(z) = \exp\{t\log\gamma(z)\}$, where the principal branch of the logarithm is chosen, has been shown to be a characteristic function for any $t \geqslant 0$ (Lukacs, p. 82). The corresponding distribution $G^{(t)}(x)$ is the Green distribution $G(x,t)$ of a homogeneous process. We now call upon DeFinetti's Theorem (Lukacs, p. 83).

Theorem II.7.1. A characteristic function $\gamma(z)$ is infinitely divisible if, and only if, it has the form

$$\gamma(z) = \lim_{m\to\infty} \exp[\nu_m\{\alpha_m(z)-1\}]$$

where $\nu_m > 0$ and the functions $\alpha_m(z)$ are characteristic functions.

In particular, the representation $\nu_m = m$ and $\alpha_m(z) = [\gamma(z)]^{1/m}$ is always available. From the theorem it follows at once that $\gamma^{(t)}(z) = \gamma(z,t) = \lim_{m\to\infty} \exp[-\nu_m t\{1-\alpha_m(z)\}]$, i.e., that every homogeneous process is the limit† of a sequence of jump processes. We next consider the Wiener-Lévy process.

(A) *The Wiener-Lévy process*

Let $X_0(t)$ be a homogeneous Markov jump process in continuous time with increment distribution $A_0(x)$ and increment frequency ν_0, for which the increments have zero mean and a finite non-zero second moment. If $X_0(0) = 0$, the characteristic function of $X_0(t)$ is $\phi_0(z,t) = \exp[-\nu_0 t\{1-\alpha_0(z)\}]$. Consider a sequence of such processes $X_{0\gamma}(t)$ with frequency $\nu_{0\gamma} = \nu_0\gamma^2$ and increment distribution $A_{0\gamma} = A_0(\gamma x)$. Then $\alpha_{0\gamma}(z) = \alpha_0(\gamma^{-1}z)$ and $X_{0\gamma}(t)$ has the characteristic function $\phi_{0\gamma}(z,t) = \exp[-\nu_0\gamma^2 t\{1-\alpha_0(\gamma^{-1}z)\}]$. Since $\alpha_0(0) = 1$ and $\alpha'_0(0) = 0$,

† See Note 2.

the exponent may be seen to be an indeterminate form as $\gamma \to \infty$. Application of L'Hôpital's rule twice then gives

$$\lim_{\gamma\to\infty} \phi_{0\gamma}(z,t) = \exp(-D z^2 t). \tag{7.1}$$

where

$$D = \frac{1}{2}\nu_0 \int x^2 dA_0(x). \tag{7.2}$$

Thus all jump processes with increments of zero mean having the same parameter D, give rise under the scaling to a common limiting process $X_D(t)$ with characteristic function $\exp(-D z^2 t)$. This process is called alternatively the *Wiener-Lévy* process, Wiener process, diffusion process, or Brownian motion process. The inverse Fourier transform of $\exp(-D z^2 t)$ is $g_D(x,t)$ of (3.23), the transition function for the Wiener-Lévy process. The Gaussian form of $g_D(x,t)$ is a manifestation of the Central Limit Theorem which we will review in Chapter III.

The sample functions of the Wiener-Lévy process are known to be continuous at all times and to have no derivative at any time. A simple and instructive argument for the continuity of the process† is available. Consider the process $X_D(t)$ with $D = 1/2$ for the time interval $0 \leqslant t \leqslant 1$, so that $g_D(x,t) = (2\pi t)^{-\frac{1}{2}} \exp\{-x^2/2t\}$. Divide the time interval (0,1) into N equal segments, each of duration N^{-1}. Let ξ_i be the increment on the ith segment. Consider the random variable $\zeta_N = \max\{|\xi_1|,|\xi_2|,\ldots |\xi_N|\}$. The distribution of ζ_N is $F_{N\zeta}(x) = P\{\zeta_N \leqslant x\}, = \prod_{i=1}^{N} P\{|\xi_i| \leqslant x\}$, from the independence of the increments ξ_i. Hence

$$F_{N\zeta}(x) = \left[1 - 2\int_{x\sqrt{N}}^{\infty} (2\pi)^{-\frac{1}{2}} e^{-y^2/2} dy\right]^N .$$

The reader will verify with the aid of L'Hôpital's rule that $\lim_{N\to\infty} F_{N\zeta}(x) = 1$ for every $x > 0$, from which the continuity of every sample function may be inferred. When the same procedure is applied to $g(x,t) = x^{t-1} e^{-x}/\Gamma(t)$ on (0,1), (cf. (3.19)), one obtains

$$F_{N\zeta}(x) = \left[1 - \left\{\Gamma\left(\frac{1}{N}\right)\right\}^{-1} \int_x^{\infty} y^{-1+N^{-1}} e^{-y} dy\right]^N .$$

Since $\Gamma(1/N) \sim N$ as $N \to \infty$, it is clear that

† For a rigorous discussion, see Doob (1953), p. 392.

$$\lim_{N\to\infty} F_{N\zeta}(x) = \exp\left\{-\int_x^\infty y^{-1}\, e^{-y}\, dy\right\},$$

and the sample functions cannot be continuous.

For the spatially homogeneous Markov process (3.20), the characteristic function is the product of the characteristic functions of the component processes, i.e.,

$$\phi(z,t) = \phi_0(z)\, e^{-\nu t\{1-a(z)\}}\, e^{-Dz^2 t}\, e^{izvt}. \tag{7.3}$$

Hence

$$\frac{\partial\phi}{\partial t}(z,t) = [-\nu\{1-\alpha(z)\} - Dz^2 + izv]\,\phi(z,t) \tag{7.4}$$

and (3.21) follows from (7.4) and inverse Fourier transformation when $Dt > 0$.

(B) *The drift processes as limiting forms*

Consider now a process $X(t) = X_0 + X_1(t) + X_2(t)$, where $X_1(t)$ and $X_2(t)$ are independent Markov jump processes. Let $X_1(t)$ be characterized by

$$\nu_{1\eta} = \eta|v|;\quad A_{1\eta}(x) = e^{\eta x} \text{ for } x \leqslant 0,\quad A_{1\eta}(x) = 1 \text{ for } x > 0, \tag{7.5}$$

and $X_2(t)$ by ν_2 and $A_2(x)$. Then $X(t)$ has the characteristic function

$$\phi_\eta(z,t) = \exp[-|v|t\{iz\eta/(\eta+iz)\} - \nu_2 t\{1-\alpha_2(z)\}]\,\phi_0(z). \tag{7.6}$$

Consider now a sequence of processes with $\eta = \eta_j$ and $\eta_j \to \infty$. Since $\phi_{\eta_j}(z,t)$ converges for every z to the function

$$\phi(z,t) = e^{-i|v|zt}\, e^{-\nu_2 t[1-a_2(z)]}\, \phi_0(z)$$

the Continuity Theorem (Lukacs, Theorem 3.6.1) states that the distribution of $X_{\eta_j}(t)$ converges to the distribution of the homogeneous drift process $X(t) = X_0 + X_2(t) - |v|t$. For processes with positive drift v, one would work with the positive exponential distributions $A_{1\eta}(x) = 1 - e^{-\eta x}$, and $\nu_{1\eta} = \eta v$ to obtain $X(t) = X_0 + X_2(t) + vt$ as a limiting form.

II.8 Some explicit Green's functions

For homogeneous processes with both positive and negative increments, the Green's functions are available in closed form for only a few simple and correspondingly important cases which we here set down.

(a) *The homogeneous diffusion process*

The process with the simplest Green's function is the homogeneous diffusion process with continuity equation (3.22). The Green density given in (3.23) is

$$g(x,t) = (4\pi Dt)^{-\frac{1}{2}} \exp\{-(x-vt)^2/4Dt\}. \tag{8.1}$$

(b) *The drift process with exponentially distributed positive increments*

For the process $X_v(t)$ of Section 3, with drift rate v, increment frequency ν, and increment distribution $A(x)$, a closed form is available when the increments are positive and exponentially distributed. When $a(x) = \lambda e^{-\lambda x}U(x)$, we have for $k \geqslant 1$,

$$a^{(k)}(x) = \frac{\lambda^k x^{k-1}}{(k-1)!} e^{-\lambda x} U(x). \tag{8.2}$$

The Green density for the process $X_v(t)$ when $v = 0$ is then given by (3.5) to be

$$g(x,t) = \delta(x)e^{-\nu t} + \left[\sum_1^\infty \{e^{-\nu t}(\nu t)^k/k!\}\{\lambda^k x^{k-1} e^{-\lambda x}/(k-1)!\}\right] U(x).$$

The power series for the *modified Bessel function* is

$$I_n(z) = \left(\frac{z}{2}\right)^n \sum_{l=0}^\infty \frac{(z^2/4)^l}{l!\,(l+n)!} \quad \text{for } n \geqslant 0; \quad I_{-n}(z) = I_n(z). \tag{8.3}$$

We have, therefore,

$$g(x,t) = \delta(x)e^{-\nu t} + \left[\lambda\nu t\, e^{-\nu t} e^{-\lambda x} \frac{I_1(2\sqrt{\lambda\nu t x})}{\sqrt{\lambda\nu t x}}\right] U(x). \tag{8.4}$$

The Green density for $X_v(t)$ is then given by

$$g_v(x,t) = g_0(x-vt,t). \tag{8.5}$$

(c) *Lattice processes with Bernoulli increments*

A homogeneous lattice process N_k will be said to have *Bernoulli increments* if the only increments which may occur are +1 and −1. Let the corresponding increment probabilities be $e_1 = p$ and $e_{-1} = q$ respectively. Then the Green's function is the transition probability $g_n^{(k)} = P\{N_k = n | N_0 = 0\}$ and is the coefficient of u^n in $\epsilon^k(u) = (q u^{-1} + pu)^k$. When $p = q = \frac{1}{2}$, we find from the Binomial theorem that $g_n^{(k)} = \beta_n^{(k)}$ where

$$\beta_n^{(k)} = \begin{cases} 2^{-k}\binom{k}{\frac{1}{2}(k-n)}, & k-n \text{ even}; \\ 0 \quad , & k-n \text{ odd}. \end{cases} \tag{8.6}$$

When $p \neq q$,

$$g_n^{(k)} = \{2^k\, q^{\frac{1}{2}(k-n)}\, p^{\frac{1}{2}(k+n)}\}\, \beta_n^{(k)}. \tag{8.7}$$

For the associated homogeneous lattice process $N(t)$ in continuous time with transitions of frequency ν, we have from (4.12) $g_n(t) = \sum_0^\infty \{e^{-\nu t}(\nu t)^k/k!\} g_n^{(k)}$. From the relation

$$\binom{k}{\frac{1}{2}(k-n)} = \frac{k!}{\{\frac{1}{2}(k-n)\}!\,\{\frac{1}{2}(k+n)\}!}$$

and the power series (8.3) for the modified Bessel function $I_n(z)$, we obtain for the symmetric case $p = q = \frac{1}{2}$,

$$g_n(t) = e^{-\nu t}\, I_n(\nu t). \tag{8.8}$$

For $p \neq q$

$$g_n(t) = (p/q)^{n/2}\, e^{-\nu t}\, I_n(\nu t \sqrt{4pq}). \tag{8.9}$$

II.9 The Green's function for lattice processes in continuous time; generalized Bessel functions

When a probability generating function $\pi(u)$ of form (6.12) is analytic at $u = 1$, its structure is such (cf. Section III.6) that it will be analytic in some annulus about $u = 0$ containing $u = 1$, and have a Laurent expansion associated therewith. If, for example, the increment generating function $\epsilon(u)$ is analytic at $u = 1$ (to be so it must have moments of all order), its Laurent expansion is the defining series $\epsilon(u) = \sum_{-\infty}^{\infty} e_n u^n$. The generating function of (6.14), $\gamma(u,t) = e^{-\nu t[1-\epsilon(u)]}$

will then also be analytic at $u = 1$, with Laurent series $\gamma(u,t) = \sum_{-\infty}^{\infty} g_n(t)u^n$. The coefficients $g_n(t)$ are given by Laurent's Theorem or (6.13) as

$$g_n(t) = \frac{1}{2\pi i}\int_C \frac{du}{u^{n+1}}\, e^{-\nu t[1-\epsilon(u)]} = e^{-\nu t} I_n^{\epsilon}(\nu t) \tag{9.1}$$

where

$$I_n^{\epsilon}(z) = \frac{1}{2\pi i}\int_C \frac{du}{u^{n+1}}\, e^{z\epsilon(u)}\,du. \tag{9.2}$$

The contour lies within the annulus and goes about the origin positively. For the symmetric single-step walk with $\epsilon(u) = \frac{1}{2}(u + u^{-1})$, (9.2) gives† $g_n(t) = e^{-\nu t} I_n(\nu t)$, where $I_n(z)$ is the modified Bessel function of order n. The functions $I_n^k(z)$ associated with $\epsilon(u) = \frac{1}{2}(u + u^{-k})$ were studied by Luchak (1956, 1957, 1958), who called them generalized Bessel functions. The more general class of functions (9.2) for arbitrary $\epsilon(u)$ (Keilson, 1963b) may also be called generalized Bessel functions by virtue of the following properties in common with $I_n(z)$:

(a) The functions $I_n^{\epsilon}(z)$ are entire and, for positive real values of z, the functions and all their derivatives are positive.‡ The power series, convergent for all z, are given by

$$I_n^{\epsilon}(z) = \sum_0^{\infty} \frac{z^m}{m!}\frac{1}{2\pi i}\int_C [\epsilon(u)]^m \frac{du}{u^{n+1}}. \tag{9.3}$$

For $[\epsilon(u)]^m$ is a probability generating function with the structure (6.12), and the coefficient of $(z^m/m!)$ in (9.3) is a probability, i.e., non-negative and less than or equal to one. Hence (9.3) is dominated by $\sum \frac{(|z|)^m}{m!} = e^{|z|}$ and is entire.

(b) The functions have a Neumann expansion

$$I_n^{\epsilon}(z+z') = \sum_{-\infty}^{\infty} I_m^{\epsilon}(z)\, I_{n-m}^{\epsilon}(z'). \tag{9.4}$$

For, from the Chapman-Kolmorogov equation, $g_n(t+t') = \sum_{-\infty}^{\infty} g_m(t) \times g_{n-m}(t')$. Equation (9.4) follows from (9.1) and analytic continuation.

---

† $e^{\frac{1}{2}z(u+u^{-1})}$ is the generating function of the modified Bessel functions. See Copson, 1935.

‡ The functions are therefore absolutely monotonic in $0 \leqslant t < \infty$. The properties of such functions are discussed in Widder (1941), Chapter IV.

(c) The functions satisfy the recursion relations

$$n\, I_n^{\epsilon}(z) \;=\; z \sum_{-\infty}^{\infty} r e_r\, I_{n-r}^{\epsilon}(z) \tag{9.5}$$

and

$$\frac{d}{dz}\, I_n^{\epsilon}(z) \;=\; \sum_{-\infty}^{\infty} e_r\, I_{n-r}^{\epsilon}(z)\,. \tag{9.6}$$

Equation (9.5) follows from $\int_C \frac{d}{du}\,\{u^{-n}\, e^{z\epsilon(u)}\}\,du = 0$, and (9.6) from differentiation under the integral sign. Equations (9.3), (9.5) and (9.6) have been given by Luchak (1956) for $\epsilon(u) = \frac{1}{2}\,(u + u^{-k})$. We observe from (9.5) for $n = 0$ that the functions I_n^{ϵ} are not linearly independent.

(d) If the structure of $\epsilon(u)$ in its annulus of convergence is sufficiently simple (for example, if $\epsilon(u)$ is "complete" as described in Section III.6), the functions $I_n^{\epsilon}(t)$ have an asymptotic dependence similar to that of the modified Bessel function. It will be shown in the next chapter that, as $t \to \infty$,

$$I_n^{\epsilon}(t) \;\sim\; \rho_0^{-n}\,(2\pi\gamma t)^{-\frac{1}{2}}\, \exp\{\epsilon(\rho_0)\,t - \tfrac{1}{2}n^2(\gamma t)^{-1}\} \tag{9.7}$$

where ρ_0 is the unique positive real zero of $\epsilon'(u)$ in its annulus of convergence, and $\gamma = \rho_0^2\, \epsilon''(\rho_0)$.

CHAPTER III

CONJUGATE TRANSFORMATION AND ASYMPTOTIC BEHAVIOUR

A Green distribution $G_k(x) = A^{(k)}(x)$ with a finite non-zero variance has a simple asymptotic behaviour as $k \to \infty$ described, for values of x sufficiently near the mean $k\mu$ of $G_k(x)$, by the familiar Central Limit Theorem. When x is substantially removed from the mean $k\mu$ so that one is dealing with "large deviations", the asymptotic behaviour of $A^{(k)}(x)$ may be obtained by a conjugate transformation technique associated with Khinchin, provided that the characteristic function $\alpha(z)$ of $A(x)$ has a convergence strip of sufficient scope. The saddlepoint approximation of Daniels for estimating $A^{(k)}(x)$ at large deviations and the Central Limit approximation associated with an optimum conjugate transformation are identical. The asymptotic properties of the Green distribution $G(x,t)$ for processes in continuous time and Green probabilities $g_n^{(k)}$ and $g_n(t)$ for lattice processes may be discussed along similar lines. The structure of the convergence strip and the properties of the conjugate family of distributions associated with this strip are examined in detail. Much of the presentation is a refinement and extension of the original paper of Daniels (1954).

III.1 The convergence strip for characteristic functions

In Section 6 of Chapter II, the principal properties on the real axis of the characteristic function †

$$\phi(z) = \int_{-\infty}^{\infty} e^{izx}\, dF(x) \tag{1.1}$$

were listed. The natural domain of the characteristic function is the real axis where the convergence of the defining integral (1.1) is assured. For most distributions of interest in applied probability theory, the integral (1.1) will converge for other values of the complex plane as well. The set of such values, i.e., the domain of convergence of (1.1), is a strip parallel to and containing the real axis. The analytic structure of the characteristic function $\phi(z)$ defined in this *convergence*

† Many authors prefer to discuss the convergence strip in terms of the two-sided Laplace Transform, for example Daniels (1954).

strip by (1.1) and defined elsewhere by analytic continuation is of basic interest. The reader is referred to Widder's *The Laplace Transform* or Lukacs' *Characteristic Functions* for a full discussion of convergence and the convergence strip. The basic results outlined in the chapter may be understood without this background.

A useful decomposition of the characteristic function may be made along the following lines. The random variable X with distribution $F(x) = P\{X \leqslant x\}$ will either be (a) positive with $F(0) = 0$; (b) non-positive with $F(0) = 1$ or (c) will have both positive and non-positive support with $0 < F(0) < 1$. In the latter case let $F_I(x) = P\{X \leqslant x | X \leqslant 0\}$ be the conditional distribution of X, given that X is non-positive, and let $F_{II}(x) = P\{X \leqslant x | X > 0\}$. Then $F(x) = q\,F_I(x) + p\,F_{II}(x)$ where $q = P\{X \leqslant 0\}$ and $p = P\{X > 0\}$, and

$$\phi(z) = q \int_{-\infty}^{0} e^{izx}\, dF_I(x) + p \int_{0}^{\infty} e^{izx}\, dF_{II}(x). \tag{1.2}$$

Let the real and imaginary parts of z be U and V respectively.

Case (a): Positive support ($q = 0$)

We then have

$$\phi(z) = \phi_{II}(z) = \int_{0}^{\infty} e^{iUx}\, e^{-Vx}\, dF_{II}(x). \tag{1.3}$$

This integral converges for all values V in the upper half-plane $V \geqslant 0$, and $\phi_{II}(z)$ will be analytic at all points in the interior of this half-plane. As $V \to +\infty$, $\phi_{II}(z) \to 0$. When V is fixed and $|U| \to \infty, \phi_{II}(z) \to 0$ if $F_{II}(x)$ is absolutely continuous for all $x > 0$; $\phi_{II}(z)$ is oscillatory if $F_{II}(x)$ is discontinuous for any $x > 0$. (Care is needed for the discussion of singular distributions; see Lukacs, pp. 26-27.)

For various distributions $F_{II}(x)$, the integral (1.3) will converge into the lower half-plane: (1) for all values $V \leqslant 0$; (2) for the values $V_{\text{min}} <$ (or $\leqslant$) $V \leqslant 0$; or (3) will converge for no negative values of V. Some simple examples illustrating each possibility follow.

(1) If the random variable X is bounded from above, i.e., if $X \leqslant a$, then $F_{II}(x) = 1$ for $x \geqslant a$, and the upper limit of integration in (1.3) may be replaced by $x = a$. $\phi_{II}(z)$ then converges for all real V, and is an entire function of z. The same conclusion follows when $F_{II}(x)$ goes to 1 quickly enough, e.g., when $1 - F_{II}(x) < \alpha e^{-\beta x^2}$ for any positive constants α, β and x sufficiently large.

(2) Suppose $F_{II}(x)$ is the *exponential distribution* $F_{II}(x) = (1 - e^{-\lambda x})\, U(x)$. Then the integral in (1.3) converges for $-\lambda < V$. The

Erlang distributions† all have such exponential tails and their characteristic functions behave in this manner. For this exponential distribution $\phi_{II}(z) = \lambda(\lambda - iz)^{-1}$ and $z = -i\lambda$ is a pole of $\phi_{II}(z)$. Note that $\phi_{II}(z)$ has a continuation into the entire z plane, even though its convergence strip ends at the singularity.

(3) If the distribution is given by

$$F_{II}(x) = (2/\pi)[\arctan(x/a)]\, U(x) \tag{1.4}$$

then (1.3) cannot converge for any negative value of V.

An ordinate of convergence $V_{\min}$ bounding the convergence strip of $\phi_{II}(z)$ will always be a singularity of $\phi_{II}(z)$. This singularity if isolated may be either a pole, an essential singularity, or a branch point. An example of the first is the Erlang distribution of order N where the singularity is a pole of order N. An essential singularity bounds the convergence strip for the probability density $f(x) = e^{-(1+x)} I_0(2\sqrt{x})\, U(x)$ where $I_0(x)$ is the modified Bessel function. This has the characteristic function (obtained from a table of Laplace transforms)

$$\phi(z) = (1-iz)^{-1} \exp\{iz(1-iz)^{-1}\}. \tag{1.5}$$

An interesting example of a branch point is that for the density

$$f_{II}(x) = F'_{II}(x) = e^2 \pi^{-\frac{1}{2}} x^{-3/2} \exp\{-x-x^{-1}\}\, U(x), \tag{1.6}$$

for which

$$\phi_{II}(z) = \exp\{2 - 2(1-iz)^{\frac{1}{2}}\}. \tag{1.7}$$

It may be noted that even though $z = -i$ is a singularity, the ordinate of convergence $V_{\min} = -i$ is part of the convergence strip. The characteristic function $\phi(z)$ has a finite limit as $z = -i$ is approached from within the convergence strip, but its derivative $\phi'_{II}(z)$ becomes infinite. When the derivative $d\phi_{II}(iV)/dV$ goes to infinity at the boundary $V_{\min}$ of its convergence strip, the convergence strip will be said to be *complete*. It is easy to construct examples where the limit of this derivative will be finite. For example, the convergence strip for the probability density $f(x) = \alpha(1+x^4)^{-1}$ ends at the real axis. But $d\phi_{II}(iV)/dV = -\alpha\int_0^\infty e^{-Vx}\, x(1+x^4)^{-1}\, dx$, and its limit is finite as V goes to zero through positive values.

---

† The *Erlang distribution of order* N has density $f_N(x) = \lambda \dfrac{(\lambda x)^{N-1}}{(N-1)!} e^{-\lambda x}\, U(x)$ and characteristic function $\phi_N(z) = [\lambda/(\lambda - iz)]^N$, where N is a positive integer.

Case (b): Non-positive support (p = 0)

Here the characteristic function is given by

$$\phi(z) = \phi_I(z) = \int_{-\infty}^{0} e^{iUx} e^{-Vx} \, dF_I(x). \qquad (1.8)$$

The integral (1.8) converges for all values V in the lower half-plane $V \leqslant 0$, and $\phi_I(z)$ is analytic at all points in its interior. Since a mass concentration at zero falls within the cognizance of $F_I(x)$, its presence will give rise to a non-zero limit for $\phi_I(z)$ when U is held fixed and $V \to -\infty$. We then have $\lim_{V\to-\infty} \phi(U+iV) = F(0) - F(0-)$. Otherwise the behaviour of $\phi_I(z)$ is in all respects identical to that of $\phi_{II}(z)$ described above. When $(d/dV)\phi_I(iV)$ becomes infinite as $V \to V_{max}$, the convergence strip for $\phi(z)$ will be said to be *complete.*

Case (c): $0 < q < 1$

When the random variable X assumes both positive and non-positive values, the integral (1.2) defining the characteristic function $\phi(z)$ will converge in the domain common to the convergence strips of $\phi_I(z)$ and $\phi_{II}(z)$. $\phi(z)$ will then be analytic in the interior of this convergence strip. This convergence strip will have zero width, i.e., reduce to the real z axis, only if both (1.3) and (1.8) do not converge beyond $V = 0$. For most of the distributions with positive and non-positive support encountered in applications, both (1.3) and (1.8) converge beyond $V = 0$ and the convergence strip contains the real z axis in its interior. The boundaries of the convergence strip for $\phi(z)$ are not in general the natural boundaries of the characteristic function which will have an analytic continuation beyond the boundaries of the strip. When the derivative of $\phi(iV)$ becomes infinite as $V \to V_{max}$ and $V \to V_{min}$ the convergence strip of $\phi(z)$ will be said to be *complete.*

A plot of $\phi(iV)$ against V in the convergence strip is shown in Fig. III.1 for two simple distributions.

III.1.1 Rouché's Theorem and the Principle of the Argument

It is often necessary to know whether a characteristic function $\phi(z)$ assumes some value A in a given region of the complex plane, and how many times it does so. Information of this kind may sometimes be obtained on the basis of Rouché's Theorem or the Principle of the Argument. These will now be stated without proof for convenience of reference, and the manner of their use indicated. The reader is referred to Copson (1935) for the proofs.

Rouché's Theorem. If $\phi(z)$ and $\psi(z)$ are two functions, regular within and on a closed contour C on which $\phi(z) \neq 0$ and on which $|\phi(z)| >$

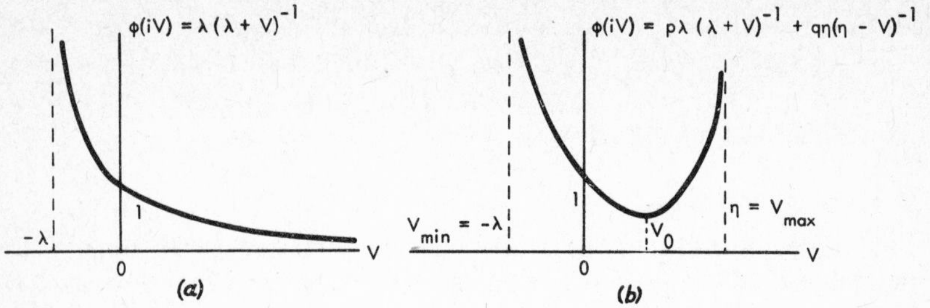

Fig. III.1

(a) $F(x) = (1 - e^{-\lambda x})\,U(x)$ (b) $F(x) = p(1 - e^{-\lambda x})\,U(x) + q e^{\eta x}\,U(-x)$

$|\psi(z)|$, then $\phi(z)$ and $\phi(z) + \psi(z)$ have the same number of zeros within C.

The Principle of the Argument. For a function $\psi(z)$ analytic within and on a contour C, the excess of the number of zeros over the number of poles of $\psi(z)$ within C, is $(1/2\pi)$ times the increase in $\arg \psi(z)$ as z goes once around C in the positive direction. A zero of multiplicity r is counted r times and a pole of order s is counted s times.

As an example, consider the function $\psi(z) = 1 - \theta(\lambda/\lambda + iz)\,\alpha^+(z)$ where $\lambda > 0$, $0 < \theta < 1$, and $\alpha^+(z)$ is the characteristic function of a distribution with positive support. Then $\alpha^+(z)$ is analytic in the upper half-plane $\text{Im}(z) > 0$, and $|\alpha^+(z)| \leqslant 1$ for $\text{Im}(z) \geqslant 0$. Consider a contour C with the form shown in Figure III.2a running from $-R_2 + i\epsilon$ to $-R_1 + i\epsilon$, along the circle of radius R_1, from $R_1 + i\epsilon$ to $R_2 + i\epsilon$, and back along

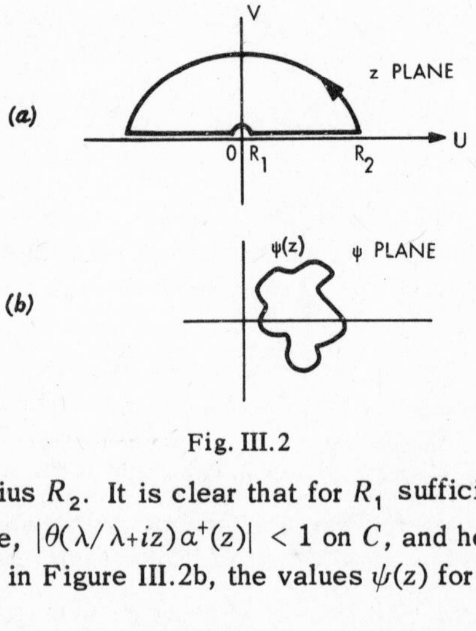

Fig. III.2

the circle of radius R_2. It is clear that for R_1 sufficiently small and R_2 sufficiently large, $|\theta(\lambda/\lambda{+}iz)\,\alpha^+(z)| < 1$ on C, and hence $\text{Re}\,\psi(z) > 0$ on C. As shown in Figure III.2b, the values $\psi(z)$ for z on C cannot encircle the origin, and $\arg\psi(z)$ must return to its original value when the contour C is traced. Since there is one simple pole for $\psi(z)$ at $z = i\lambda$, there must be exactly one simple zero inside C. If we let R_1 go to zero and R_2 go to infinity, we infer that $\psi(z)$ has precisely one simple zero in the half-plane $\mathrm{Im}\,(z) \geqslant 0$.

III.2 The conjugate family of distributions†

For every distribution $F_0(x)$ whose characteristic function $\phi_0(z)$ has a convergence strip of finite width, there is a simple family of distributions related to $F_0(x)$ via its convergence strip. For let $z = iV$ belong to the convergence strip. Then $\phi_0(iV)$ is real and positive and

$$F_V(x) = \{\phi_0(iV)\}^{-1} \int_{-\infty}^{x} e^{-Vx'}\, dF_0(x') \tag{2.1}$$

is a probability distribution. The set of distributions $\{F_V(x)\}$ for all values V in the convergence interval (V_{min}, V_{max}) will be called the *conjugate family of distributions*. From (2.1), the characteristic function of $F_V(x)$ is

$$\phi_V(z) = \phi_0(z+iV)/\phi_0(iV). \tag{2.1a}$$

We note that $\phi_0(U+iV) = \phi_0(iV)\,\phi_V(U)$. This implies that the function $\phi_0(U+iV)$ is itself a characteristic function multiplied by a positive constant $\phi_0(iV)$. We also note that for other than the degenerate distribution $F_0(x) = U(x)$, $\phi_0(iV)$ is a convex‡ function of V in the interior of the convergence strip, i.e.,

$$\frac{d^2}{dV^2}\,\phi_0(iV) = \int_{-\infty}^{\infty} x^2\, e^{-Vx}\, dF_0(x) > 0\,. \tag{2.1b}$$

The mean of the distribution $F_V(x)$, obtained by differentiating (2.1a), is given by

$$\mu_V = \int_{-\infty}^{\infty} x\, dF_V(x) = -\frac{\dfrac{d}{dV}\phi_0(iV)}{\phi_0(iV)} = -\frac{d}{dV}\log\phi_0(iV)\,, \tag{2.2}$$

and its variance by

---

† Conjugate distributions are believed to have been introduced by Esscher (1932), and have been employed subsequently by Cramér ((1938), Wald (1945), Khinchin (1949) and others. These references are due to E.J. Williams (unpublished report).

‡ It is also superconvex. See Note 3.

$$\sigma_V^2 = E(X_V^2) - \mu_V^2 = \frac{\frac{d^2}{dV^2}\phi_0(iV)}{\phi_0(iV)} - \left\{\frac{\frac{d}{dV}\phi_0(iV)}{\phi_0(iV)}\right\}^2. \tag{2.3}$$

If we differentiate (2.2), we find by comparison with (2.3) the important result

$$\frac{d\mu_V}{dV} = -\sigma_V^2. \tag{2.4}$$

The variance σ_V^2 of $F_V(x)$ will be positive (apart from the degenerate distributions $U(x-v)$, which can never be conjugate to non-degenerate distributions). We then have the following theorem.

Theorem III.2.1. For a non-degenerate conjugate family of distributions $F_V(x)$ the mean μ_V of $F_V(x)$ is a continuous monotonic decreasing function of V.

The range of values assumed by μ_V for the conjugate family of distributions is of interest. To discuss this range, we introduce the greatest lower bound x_L and least upper bound x_M of the random variable X of distribution $F_0(x)$, defined by

$$x_L = \sup\{x: F_0(x) = 0\}$$

and (2.5)

$$x_M = \inf\{x: F_0(x) = 1\}.$$

These values may be infinite.

Theorem III.2.2. If the distribution $F_0(x)$ has a complete (see Section 1) convergence strip, then for each value α for which $x_L < \alpha < x_M$, there will be one member $F_\theta(x)$ of the conjugate family of distributions with mean $\mu_\theta = \alpha$.

The proof is an extension of an argument given by Daniels. By definition of x_L and x_M, we have $0 < F_0(\alpha-) \leqslant F_0(\alpha+) < 1$. Let $B_\alpha(x) = F_0(x+\alpha)$. Then

$$\beta_\alpha(z) = e^{-iz\alpha}\phi_0(z), \tag{2.6}$$

the c.f. of $B_\alpha(x)$, has the same convergence strip as $\phi_0(z)$, and

$$\zeta_\alpha(V) = -\frac{d}{dV}\beta_\alpha(iV) = \int_{-\infty}^{\infty} x\, e^{-Vx}\, dB_\alpha(x) \tag{2.7}$$

holds in that convergence strip. Since $0 < B_\alpha(0-) \leqslant B_\alpha(0+) < 1$, the distribution $B_\alpha(x)$ has both positive and negative support. Consequently if

$V_{max} = +\infty$, the behaviour of the integral in (2.7) is such that $\zeta_\alpha(+\infty) = -\infty$, and if $V_{min} = -\infty$, $\zeta_\alpha(-\infty) = +\infty$. If V_{max} is finite, from the completeness assumed, $\zeta_0(iV)$ and $\zeta_\alpha(iV)$ both approach $-\infty$ when $V \to V_{max}$. Similarly when V_{min} is finite, $\zeta_0(iV)$ and $\zeta_\alpha(iV)$ both approach $+\infty$. But from (2.7)

$$\frac{d\zeta_\alpha(V)}{dV} = -\int_{-\infty}^{\infty} x^2\, e^{-Vx}\, dF_\alpha(x) < 0 \tag{2.8}$$

and $\zeta_\alpha(V)$ is a continuous monotonic decreasing function of V. Hence the completeness assures that $\zeta_\alpha(V)$ takes on all real values and in particular the value zero for some value θ inside the convergence interval. Since $[(d/dV)(\phi_0(iV)e^{\alpha V})]_\theta = 0$ implies from (2.2) that $\mu_\theta = \alpha$, the theorem is proved.

When the convergence strip is not complete, we can only infer that μ_V takes on the value α when

$$\mu_{max} < \alpha < \mu_{min}, \tag{2.9}$$

where

$$\mu_{max} = \lim_{V \to V_{max}^-} \frac{-\frac{d}{dV}\phi_0(iV)}{\phi_0(iV)} \tag{2.10a}$$

and

$$\mu_{min} = \lim_{V \to V_{min}^+} \frac{-\frac{d}{dV}\phi_0(iV)}{\phi_0(iV)} \tag{2.10b}$$

The existence of a minimum for $\phi(iV)$ in the interior of the strip corresponds by (2.2) to the availability of a conjugate distribution with zero mean. The presence of such a distribution when $x_L < 0 < x_M$ and the convergence strip is complete is assured by Theorem 2.2, as stated in the following corollary.

Corollary III.2.2. If a distribution $F_0(x)$ has both positive and negative support and if its convergence strip is complete, $\phi(iV)$ assumes a minimum inside the convergence strip.

The behaviour of μ_V and $\phi(iV)$ for a typical conjugate family of distributions with a complete convergence strip is shown in Fig. III.3. Here $V_{max} = +\infty$, x_L is finite and negative, and $x_M = +\infty$. The completeness is evident in the infinite slope of $\phi(iV)$ at V_{min} and V_{max}.

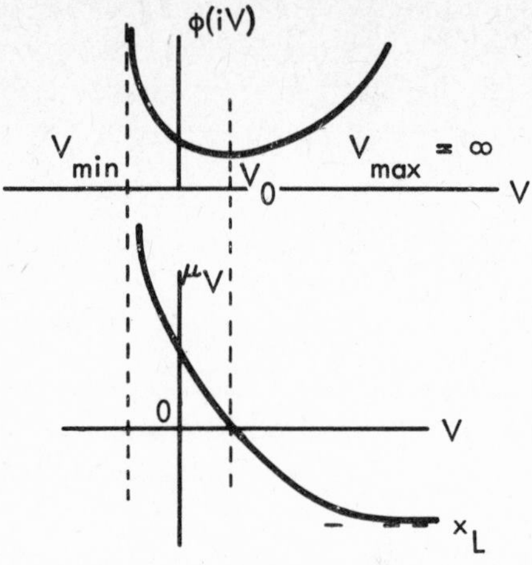

Fig. III.3

The convex function $\phi(iV)$ takes its minimum at V_0 where the corresponding value of μ_V is zero.

Simple examples of conjugate families of distributions are easily given.

(1) If $f_0(x)$ is the Gaussian density $(2\pi)^{-\frac{1}{2}}\, e^{-\frac{1}{2}x^2}$, then $f_V(x) = (2\pi)^{-\frac{1}{2}}\, e^{-\frac{1}{2}(x+V)^2}$, and the convergence interval is $(-\infty, \infty)$. The mean $\mu_V = -V$ assumes all real values and σ_V^2 is constant over the convergence interval.

(2) For the Binomial Distribution $F_0(x) = \sum_{r=0}^{n} \binom{n}{r} p_0^r\, q_0^{n-r}\, U(x-r)$ of index n with $0 < p_0 < 1$ it is readily seen that the convergence interval is $(-\infty, \infty)$. All distributions in the family are Binomial with the same index n and $F_V(x)$ has $p_V = (p_0\, e^{-V} + q_0)^{-1}\, p_0\, e^{-V}$. Thus μ_V assumes all values in $(0,n)$, σ_V is bounded, and the limiting distributions $U(x)$ and $U(x-n)$ are approached as $V \to \pm\infty$.

(3) If $F_0(x)$ is the Poisson Distribution $\sum_0^\infty \{\lambda^n\, e^{-\lambda}/n!\}\, U(x-n)$ with parameter λ, $F_V(x)$ is the Poisson distribution with parameter $\lambda\, e^{-V}$, for V on $(-\infty, \infty)$.

(4) The Gamma Distribution with density $f_0(x;\kappa,\lambda) = \{\Gamma(\kappa)\}^{-1} \times \lambda^\kappa\, x^{\kappa-1}\, e^{-\lambda x}\, U(x)$ has conjugate Gamma Distributions with $f_V(x;\kappa,\lambda) = f_0(x;\kappa,\lambda+V)$.

III.3 The saddlepoint employed in approximating sums of independent random variables

Saddlepoint methods for approximating integrals arise in connection with the Central Limit Theorem in the following way. The distribution of the sum X_N of N independent identically distributed random variables of distribution $F(x)$ is the convolution $F^{(N)}(x)$ (II.3.2) and its characteristic function is $\phi^N(z)$. If $F(x)$ is absolutely continuous and its density $f(x)$ is suitably behaved on $(-\infty,\infty)$, the simple inversion formula (II.6.5) may be employed. The following lemmas will be needed subsequently.

Lemma III.3.1. If $F(x)$ is absolutely continuous and its density $f(x)$ is of finite total variation A on $(-\infty,\infty)$, then when z is real,†

$$|\phi(z)| < A|z|^{-1}. \tag{3.0}$$

For $\phi(z) = \int_{-\infty}^{\infty} e^{izx} f(x)\,dx = (iz)^{-1}\int_{-\infty}^{\infty} f(x)\,d(e^{izx})$. This may be regarded as a Stieltjes integral and integrated by parts. Because $f(x)$ is integrable and of bounded variation, $f(x) \to 0$ as $x \to \pm\infty$, and there is no contribution from the end terms. Thus $\phi(z) = iz^{-1}\int_{-\infty}^{\infty} e^{izx}\,df(x)$, and $|\phi(z)| < |z|^{-1}\int|df(x)| = A|z|^{-1}$, as stated.

Lemma III.3.2. If $F(x)$ is absolutely continuous, and $f^{(m)}(x)$ is bounded for some m, then for all $N \geqslant 2m$, $f^{(N)}(x)$ is continuous and

$$f^{(N)}(x) = \frac{1}{2\pi}\int_{-\infty}^{\infty} \phi^N(z)\,e^{-izx}\,dz. \tag{3.1}$$

For, if $f^{(m)}(x) \leqslant f_{max}$, $\int\{f^{(m)}(x)\}^2 dx \leqslant f^{(m)}_{max}\int f^{(m)}(x)\,dx = f_{max}$. Moreover, by Parseval's theorem for L_2 functions (Titchmarsh, 1949, Chapter III), $\int\{f^{(m)}(x)\}^2\,dx = (2\pi)^{-1}\int|\phi(U)|^{2m}\,dU < \infty$, and the lemma follows from (II.6.5).

Note that if $\int\{f(x)\}^{1+\delta}\,dx < \infty$ for some $0 < \delta < 1$, (3.1) will be valid. For then $\phi(U) \in L_r(-\infty,\infty)$ for $r = (1+\delta)/\delta$ (Titchmarsh, Theorem 74) and for $m \geqslant r$, $\phi^m(U)$ is integrable and $f^{(m)}(x)$ is bounded.

---

†Lemma III.3.1′. If $f(x)$ has total variation A on $(-\infty,\infty)$, and positive support only, then $|\phi(z)| < A|z|^{-1}$ for all z in the upper half-plane $V \geqslant 0$. The argument required is a trivial modification of that for Lemma III.3.1.

The integrand of (3.1) may be written in the form $e^{N[\log \phi(z) - izx/N]}$, where, since $\phi(iV) > 0$, the exponent is analytic at every point of the imaginary axis $z = iV$ in the interior of the convergence strip. One is interested in large values of N. If one can find a convenient point z_0 available at which the derivative of the exponent vanishes, one has the classical setting for the *saddlepoint method.*

A related saddlepoint expansion has been given by Daniels (1954). A discussion of saddlepoint expansion will be given in Section 5(B). For the moment we require only the definition of a saddlepoint. A function $e^{N\psi(z)}$ analytic in a neighbourhood of z_0 is said to have a saddlepoint at $z = z_0$, if $\psi'(z_0) = 0$, i.e., if $[(d/dz)\, e^{N\psi(z)}]_{z_0} = 0$. Hence the integrand of (3.1) has a saddlepoint at z when $\phi'(z)/\phi(z) = ix/N$. For differentiation in the V direction this becomes

$$\frac{-\dfrac{d}{dV}\,\phi(iV)}{\phi(iV)} = \mu_V = x/N\,. \tag{3.2}$$

The availability of a saddlepoint may now be inferred from Theorem 2.2.

Theorem III.3.3. Let $F(x)$ be absolutely continuous and let its density $f(x)$ at a point of continuity of $f(x)$ be given by (3.1). If the convergence strip is complete, then the integrand of (3.1) will have one and only one saddlepoint on the imaginary axis in the convergence strip when

$$x_L < x/N < x_M\,. \tag{3.3}$$

The bounds x_L and x_M are defined by (2.5).

The saddlepoint contour is the line $z = iV + U$ through the saddlepoint parallel to the real axis.

III.4.1 Conjugate transformation and the Central Limit Theorem

Consider the random variable X_0 with distribution $F_0(x)$. If X_0 has mean μ_0 and standard deviation σ_0, the sum $X_N = \sum_1^N X_{0i}$ of N such independent random variables with distribution $F_{0N}(x) = F_0^{(N)}(x)$ has mean $N\mu_0$ and standard deviation $N^{\frac{1}{2}}\sigma_0$. If one were interested, for a particular value of x and large N, in the probability $P\{x - \frac{1}{2}\Delta < X_N \leqslant x + \frac{1}{2}\Delta\} = F_{0N}(x + \frac{1}{2}\Delta) - F_{0N}(x - \frac{1}{2}\Delta)$ and wished to employ the Central Limit Theorem directly, one might find the ratio $(x - N\mu_0)/(N^{2/3}\sigma_0)$ discouragingly high. (A criterion for the validity of the Central Limit approximation will be given in Section 5.1.) A more efficient use of the Central Limit Theorem is possible when a conjugate family of distributions (Section 2) is available, as we will now see.

Suppose for simplicity that the conjugate family $\{F_V(x)\}$ is complete (cf. Theorem 2.2). For given values of x and N, such that $Nx_L < x < Nx_M$, there will be according to Theorem 2.2 precisely one member of the family, $F_\theta(x)$, for which the mean μ_θ has the value

$$\mu_\theta = (x/N). \tag{4.1}$$

For the sum of N independent random variables with this distribution the ratio $(x - N\mu_\theta)/(N^{2/3}\sigma_\theta)$ is zero, and in this sense, the distribution $F_\theta(x)$ is the member of the conjugate family $\{F_V(x)\}$ from which optimum results for given x and N may be expected from the Central Limit Theorem. Transformation to and from this distribution is exact. The distribution $F_\theta(x)$ will be said to be the *optimum* distribution for a given value of x and N. From (3.2) we see that the saddlepoint of (3.1), is $z = i\theta$ where θ is the parameter for the optimum distribution. The saddlepoint approximation defined below and the Central Limit approximation for the optimum distribution are identical. The latter is somewhat more general since it permits discussion of distributions which are not absolutely continuous.

From (2.1a) one has $\phi_0^N(z) = \phi_0^N(iV)\,\phi_V^N(z - iV)$. For $V = \theta$, it follows that

$$F_0^{(N)}(x) = \phi_0^N(i\theta) \int_{-\infty}^{x} e^{\theta x'}\, dF_\theta^{(N)}(x') \tag{4.2}$$

where $F_\theta(x)$ is given by (2.1). If $F_0(x)$ is absolutely continuous, all members $F_V(x)$ of the conjugate family of distributions and their N-fold convolutions are also absolutely continuous, and we have in place of (4.2)

$$f_0^{(N)}(x) = \phi_0^N(i\theta) e^{\theta x} f_\theta^{(N)}(x). \tag{4.3}$$

All the apparatus of the Central Limit Theorem may now be applied to (4.2) and (4.3).

The function $f_\theta^{(N)}(x) = f_\theta^{(N)}(N\mu_\theta)$ describes the N-fold composition of a probability density at its mean. Under very broad conditions, it will be seen that for x fixed,

$$(2\pi N\sigma_\theta^2)^{\frac{1}{2}} f_\theta^{(N)}(N\mu_\theta) \to 1 \qquad \text{as } N \to \infty. \tag{4.4}$$

With $f_\theta^{(N)}(x) \sim (2\pi N\sigma_\theta^2)^{-\frac{1}{2}}$, (4.3) becomes the asymptotic equation†

---

† See footnote on next page.

$$f_0^{(N)}(x) \sim \phi_0^N(i\theta)\, e^{\theta x} (2\pi N\sigma_\theta^2)^{-\frac{1}{2}} = g_N(x) \qquad \text{as } N \to \infty. \qquad (4.5a)$$

The "*saddlepoint approximation*" $f_0^{(N)}(x) \approx g_N(x)$ implied by (4.5a) for N sufficiently large, was introduced to statistics by Daniels (1954) in a different context (see Section III.5.2). The relation (4.4) on which the asymptotic relation (4.5a) rests will be valid under the following typical conditions:

(C_1) If (A) $f_0(x)$ has both positive and negative support; (B) $\phi_0(iV)$ has a minimum at a point V_0 (Fig. III.1b) inside the convergence interval (as in Corollary III.2.2); (C) $f_0^{(m)}(x)$ is bounded for some finite m; then (4.4) and (4.5a) are valid.

For $\mu_\theta = x/N \to 0$, when $N \to \infty$ and $\theta \to V_0$. We note for $m = 1$ that $f(x) < f_{\max}$ implies that $\{f(x)\}^{1+\delta}$ is integrable for $\delta > 0$, since $\{f(x)\}^{1+\delta} < f_{\max}^{\delta} f(x)$. Moreover $\{f_V(x)\}^{1+\delta}$ will be integrable for some $\delta_V > 0$ for every value V in the interior of the convergence interval, since $\{f(x)e^{-Vx}\}^{1+\delta} < f_{\max}^{\delta}\{f(x)e^{-V(1+\delta)x}\}$. Identical conclusions follow for $m > 1$. The local form of the Central Limit Theorem (Gnedenko and Kolmogorov (1954), Section 46, Theorem 1) then states that

$$\lim_{N\to\infty} (N\sigma_\theta^2)^{\frac{1}{2}} f_\theta^{(N)}(N\mu_\theta + N^{\frac{1}{2}}\sigma_\theta x) = (2\pi)^{-\frac{1}{2}} e^{-\frac{1}{2}x^2}.$$

From the convergence of $f_\theta(x)$ to $f_{V_0}(x)$ and the uniformity associated with V_0 being inside the convergence interval, (4.4) and (4.5a) may be anticipated. A proof will be given in Section 5.1 and Note 4.

It will be seen in Section 5.1 that one also has

$$f_0^{(N)}(x) \sim \phi_0^N(iV_0)\, e^{V_0 x} (2\pi N\sigma_{V_0}^2)^{-\frac{1}{2}} \qquad \text{as } N \to \infty. \qquad (4.5b)$$

The approximation implicit in (4.5b) is in general poorer than the saddlepoint approximation implicit in (4.5a). It is, however simpler in structure, since V_0 is independent of x, and correspondingly useful. It will be shown in Section 5.1 that the approximation in (4.5b) is valid when $|x^2/2N\sigma_{V_0}^2| \ll 1$.

---

† (See previous page.) The notation $\sim$ is employed for two functions to imply that their ratio has the limit one, i.e., $\psi_N \sim \phi_N$, as $N \to \infty$, implies that $\lim_{N\to\infty} \psi_N/\phi_N = 1$. In particular, if, for two infinitesimals $\psi_N \sim \phi_N$ then $\psi_N - \phi_N$ is a higher order infinitesimal than ϕ_N. When $\psi_N \sim \phi_N$, ϕ_N is said to be an asymptotic representation of ψ_N.

(C_2) If $f_0(x)$ is bounded and has positive support only, and if $f_0(x) \sim \beta x^\alpha$ as $x \to 0+$, where $\beta > 0$, $\alpha \geqslant 0$, then (4.4) and (4.5a) are valid.

Here again $\mu_\theta = x/N \to 0$ as $N \to \infty$, but $\phi_0(iV)$ does not have a minimum (Fig. III.1a), and $\theta \to +\infty$ as $N \to \infty$. Nevertheless $f_V(x)$ after standardization to zero mean and unit variance will have a limiting form as $V \to \infty$, and this is all that is required for the validity of (4.4), as we will see in III.4.2.

The variance σ_θ^2 required for (4.5a) and (4.5b) may be obtained from (2.3) and is given by

$$\sigma_\theta^2 = -(x/N)^2 + [\phi_0(i\theta)]^{-1} \int_{-\infty}^{\infty} x^2 e^{-\theta x} dF_0(x). \tag{4.5'}$$

If $F_0(x)$ is not absolutely continuous, a generalization of (4.5b) is available based on the integral form of the Central Limit Theorem. From (4.2) we may write

$$F_0^{(N)}(x+\tfrac{1}{2}\Delta) - F_0^{(N)}(x-\tfrac{1}{2}\Delta) = \phi_0^N(i\theta) \int_{x-\Delta/2}^{x+\Delta/2} e^{\theta x'} dF_\theta^{(N)}(x'). \tag{4.6}$$

Now let $\Delta = N^{\frac{1}{2}}\sigma_\theta\delta$ and $x' = x + N^{\frac{1}{2}}\sigma_\theta y$ in (4.6). Again for x fixed and $N \to \infty$, the parameter $\theta \to V_0$, the minimum of $\phi_0(iV)$ in its strip. By virtue of the integral form of the Central Limit Theorem, we have the asymptotic equation

$$F_\theta^{(N)}(x+N^{\frac{1}{2}}\sigma_\theta y) \sim \int_{-\infty}^{y} (2\pi)^{-\frac{1}{2}} \exp\{-\tfrac{1}{2}y'^2\}dy' \quad \text{as } N \to \infty, \tag{4.6'}$$

where $\mu_\theta = x/N$. Consequently, we have for x and δ fixed as $N \to \infty$,

$$F_0^{(N)}(x+\tfrac{1}{2}\delta N^{\frac{1}{2}}\sigma_\theta) - F_0^{(N)}(x-\tfrac{1}{2}\delta N^{\frac{1}{2}}\sigma_\theta) \tag{4.7}$$

$$\sim \frac{e^{\theta x}\,\phi_0^N(i\theta)}{\sqrt{2\pi}} \int_{-\delta/2}^{\delta/2} \exp\{-\tfrac{1}{2}y^2 + \theta N^{\frac{1}{2}}\sigma_\theta y\}dy\,.$$

The integral may be re-expressed in terms of the error function for computation purposes.

The evaluation of θ as a function of x/N required for (4.5a) and (4.7) is straightforward. The function will, in simple instances, be available explicitly. More generally, θ will be obtained by a numerical plot of μ_θ against θ, with μ_θ obtained from (2.2). One of several examples of the accuracy of the saddlepoint approximation given by Daniels is the following.

(1) For the Gamma distribution with density $f_0(x;\kappa,\lambda) = \{\Gamma(\kappa)\}^{-1} \times \lambda^\kappa x^{\kappa-1} e^{-\lambda x} U(x)$, $\phi_0(z) = \{\lambda/(\lambda - iz)\}^\kappa$ and from (2.2) $\mu_V = \kappa(\lambda + V)^{-1}$. Hence $\theta = \kappa(N/x) - \lambda$. From (2.4), $\sigma_\theta = \kappa^{-\frac{1}{2}}(x/N)$. The saddlepoint approximation (4.5a) then says that

$$f_0^{(N)}(x) \approx (\kappa N)^{1-N\kappa}(2\pi N\kappa)^{-\frac{1}{2}} e^{N\kappa} \lambda^{N\kappa} x^{N\kappa-1} e^{-\lambda x}. \tag{4.8}$$

The exact result is $f_0^{(N)}(x) = f_0(x, N\kappa, \lambda)$ and this differs from the saddlepoint approximation (4.8) only in that $\Gamma(N\kappa)$ has been replaced by its Stirling approximation.

It should be noted that $F_0(x)$ need not have any moments for the saddlepoint approximation to be available. For if $F_0(x)$ is the distribution of a positive random variable, the conjugate distributions $F_V(x)$ for $V > 0$ all exist and have all moments. The availability of an optimum distribution is assured by Theorem 2.2. This corresponds to the analyticity of the characteristic function $\phi_0(z)$ in the upper half-plane, even if $\phi_0(z)$ is not well behaved at $z = 0$.

Values of x for given N lying outside the interval in which the central limit approximation is valid are known as large deviations. Some basic papers discussing large deviations for sums of identically distributed random variables are those of Cramér (1938), Daniels (1954), and Richter (1957). Two important references on the Central Limit Theorem are the books of Gnedenko and Kolmogorov (1954), and Cramér (1962).

III.4.2 Uniformity of the saddlepoint approximation

The uniformity of the saddlepoint approximation (4.5a) over the range of values (Nx_L, Nx_M) assumed by x_N has been demonstrated under broad conditions by Daniels. This uniformity has a simple interpretation in terms of the conjugate distributions and the Central Limit Theorem. For the absolutely continuous case, for example, the saddlepoint approximation $f_0^{(N)}(x) \approx g_N(x)$ contained in (4.5a) is based on the exact equation (4.3) and the central limit approximation of the local density $f_\theta^{(N)}(x)$ about the mean of $X_{\theta N}$. The accuracy of this latter approximation depends only on the shape of $f_\theta(x)$, i.e., on the shape of the standardized density $f_{s\theta}(x) = \sigma_\theta f_\theta(\mu_\theta + \sigma_\theta x)$ with zero mean and unit variance. For many of the distributions encountered in practice, the standardized density $f_{sV}(x)$ approaches a limiting density of Gamma or normal type, as V approaches the boundaries of the convergence interval. Consequently, the relative error $\{1-g_N(x)/f_0^{(N)}(x)\}$ of the saddlepoint approximation has a limiting value as well, and the uniformity is thereby assured. Two simple examples given by Daniels will illustrate this limiting behaviour of $f_{SV}(x)$.

(A) Suppose that X_0 is a positive random variable having a distribution $F_0(x)$ differentiable in some neighbourhood of the origin, and that the local density $f_0(x) = F_0'(x)$ has the asymptotic behaviour

$$f_0(x) \sim \beta x^{\alpha} \qquad \text{as } x \to 0+; \qquad \beta > 0,\ \alpha \geqslant 0.$$

The convergence strip of $\phi_0(z)$ contains the half-plane $V > 0$. For large V, the conjugate distribution $F_V(x)$ has its support localized about its origin, and $f_V(x) \sim \beta e^{-Vx} x^{\alpha}/\phi_0(iV)$. The limiting form of the standardized $F_{SV}(x)$ will be the Gamma distribution $\int_{-\infty}^{x} y^{\alpha} e^{-y}\, dy/\Gamma(\alpha+1)$ standardized to zero mean and unit variance.

(B) If for the random variable X_0,

$$f_0(x) \sim A\exp(-\beta x^{\alpha}) \qquad \text{as } x \to \infty; \qquad \beta > 0,\ \alpha > 1,$$

then the limit of $F_{SV}(x)$ as $V \to -\infty$ is the normal distribution. For $f_V(x) \sim \psi_V(x) = A\exp(-\beta x^{\alpha} - Vx)/\phi_0(iV)$ as $x \to \infty$, and $\psi_V(x)$ has a maximum at $x_0 = (-V/\alpha\beta)^{1/(\alpha-1)}$. If we set $x = x_0 y$, we have $\psi_V(x) \approx \psi_V(x_0) \times \exp[-\frac{1}{2}\beta\,\alpha(\alpha-1)\, x_0^{\alpha}(y-1)^2]$, from which the limiting normal form of $F_{SV}(x)$ is clear.

III.5.1 The domain of validity of the central limit approximation; zones of convergence

The saddlepoint approximation gives rise to a coarse but simple criterion for estimating the range of values (x,N) for which the local *central limit approximation*†

$$f_0^{(N)}(x) \approx (2\pi N\sigma_0^2)^{-\frac{1}{2}} \exp\{-(x-N\mu_0)^2/(2N\sigma_0^2)\} \tag{5.1}$$

may be expected to be valid. The rule states that if the departure of $f_0(x)$ from normality is not excessive, one must satisfy the two conditions:

$$\text{(a)}\quad N \gg 1; \qquad \text{and} \qquad \text{(b)}\quad \frac{|x - N\mu_0|}{N^{2/3}\sigma_0} \ll 1. \tag{5.2}$$

† A more precise asymptotic relation is given in Note 4.

To see this, consider a pair of values of x and N and let $\zeta = (x - N\mu_0)/N$. Let $i\theta$ be the saddlepoint for (x/N). The saddlepoint equation (3.2) may be written in the form

$$H(\theta) = \frac{\int x\, e^{-\theta x} f_0(x)\, dx}{\int e^{-\theta x} f_0(x)\, dx} = \frac{x}{N} = \mu_0 + \zeta . \tag{5.3}$$

If we introduce the standardized density $f_{s0}(y) = \sigma_0 f_0(\mu_0 + \sigma_0 y)$, equation (5.3) becomes

$$H_s(\theta\sigma_0) = \frac{\int y\, e^{-\theta\sigma_0 y} f_{s0}(y)\, dy}{\int e^{-\theta\sigma_0 y} f_{s0}(y)\, dy} = \frac{\zeta}{\sigma_0}. \tag{5.4}$$

The function $H_s(Z)$ is an analytic function of $Z = \theta\sigma_0$ about $Z = 0$, with $H_s(0) = 0$ and $H_s'(0) = -1$. Hence (5.4) defines Z as an inverse analytic function of ζ/σ_0 with the Taylor expansion

$$Z = \sigma_0\theta = -(\zeta/\sigma_0) + K_2(\zeta/\sigma_0)^2 + K_3(\zeta/\sigma_0)^3 , \tag{5.5}$$

where the coefficients K_j are dimensionless by virtue of the standardization. If the departure of the standardized distribution from normality is not excessive, these coefficients will be comparable to unity in magnitude. Consequently, if $|\zeta/\sigma_0| \ll 1$, then $|\sigma_0\theta| \ll 1$, and

$$\theta \approx -\zeta/\sigma_0^2 . \tag{5.5a}$$

A simple example is provided by the exponential density $f_0(x) = \lambda e^{-\lambda x} \times U(x)$, for which $\mu_0 = \lambda^{-1}$ and $\sigma_0 = \lambda^{-1}$. From (5.3) we have $(\lambda + \theta)^{-1} = \lambda^{-1} + \zeta$, so that $(1+Z)^{-1} = 1 + w$ where $w = \zeta/\sigma_0$. Hence $Z = -w(1+w)^{-1}$ and $|Z| \ll 1$ when $|w| \ll 1$. For the Erlang density $f_0(x) = \{\lambda(\lambda x)^{N-1} e^{\lambda x}/(N-1)!\}U(x)$, we find in the same way, $Z = -w[1+N^{-\frac{1}{2}}w]^{-1}$. Thus the radius of convergence of the series (5.5) becomes infinite as $N \to \infty$ and $f_{s0}(x)$ approaches the normal distribution. For any Gaussian $f_0(x)$, $Z = -w$ is exact.

We now take the saddlepoint approximation (4.5a) and rewrite it as

$$f_0^{(N)}(x) \approx \frac{\exp\{N \log \phi_0(i\theta) + \theta x\}}{\sqrt{2\pi N \sigma_\theta^2}} . \tag{5.6}$$

When $|\zeta/\sigma_0|$ is small, θ will be close to zero and the behaviour of the denominator will be adequately represented by replacing σ_θ^2 by σ_0^2. The rapid variations occur in the exponent. When this is expanded about $\theta = 0$ we have $N \log \phi_0(i\theta) + \theta x \approx (-N\mu_0 + x)\theta + \frac{1}{2} N \sigma_0^2 \theta^2$ to terms quadratic in $\sigma_0 \theta$. But $x - N\mu_0 = N\zeta$ and $\theta \approx -\zeta/\sigma_0^2$. Hence $N \log \phi_0(i\theta) + \theta x \approx -\frac{1}{2} N\zeta^2/\sigma_0^2 = -\frac{1}{2}(x - N\mu_0)^2/N\sigma_0^2$ giving (5.1). Under condition (5.2b), higher order terms in $\sigma_0\theta$ will be negligible (see Note 4). Condition (5.2a) is required for the saddlepoint approximation (5.6).

The requirement that the departure of the initial distribution from normality be not excessive is better understood with the help of an example. Suppose $F_0(x) = (1-q)\,U(x) + q\,A(x)$ where $A(x)$ is absolutely continuous and has zero mean. The distribution $F_0^{(N)}(x)$ has a probability mass $(1-q)^N$ concentrated at $x = 0$. If q is very small, N must be correspondingly large before this mass is dissipated and there can be hope of accuracy from the central limit approximation. Normality will of course ultimately set in.

A family of similar Gaussian approximations may be exhibited in the same manner. Thus suppose one is interested in a sequence of values (x_N, N) where x_N/N is clustered about some value μ_V. Let $x_N/N = \mu_V + \zeta_N$ and let θ_N be the saddlepoint for x_N/N. The saddlepoint equation (5.3) may then be written in the form (omitting the subscripts N),

$$H(\theta) = \frac{\int x \exp\{-(\theta - V)x\}\, f_V(x)\, dx}{\int \exp\{-(\theta - V)x\}\, f_V(x)\, dx} = \mu_V + \zeta. \tag{5.7}$$

For the standardized density $f_{SV}(y) = \sigma_V f_V(\mu_V + \sigma_V y)$, (5.4) becomes

$$\frac{\int y \exp\{-(\theta - V)\sigma_V y\}\, f_{SV}(y)\, dy}{\int \exp\{-(\theta - V)\sigma_V y\}\, f_{SV}(y)\, dy} = \frac{\zeta}{\sigma_V}. \tag{5.8}$$

Clearly $\theta - V \approx -\zeta/\sigma_V^2$ by the same reasoning as above. From the saddlepoint approximation (4.5a) we then have

$$f_0^{(N)}(x) = \frac{\phi_0^N(iV)\, e^{Vx}}{\sqrt{2\pi N \sigma_\theta^2}} \exp\{N \log \phi_0(i\theta) - N \log \phi_0(iV) + x(\theta - V)\}. \tag{5.9}$$

Proceeding as above, we then find that

$$f_0^{(N)}(x) \approx \frac{\phi_0^N(iV)\, e^{Vx}}{\sqrt{2\pi N \sigma_V^2}} \exp\{-(x - N\mu_V)^2/2N\sigma_V^2\}, \tag{5.10}$$

provided that:

$$\text{(a)}\ N \gg 1; \quad \text{and} \quad \text{(b)}\ \frac{|x - N\mu_V|}{N^{2/3}\sigma_V} \ll 1. \qquad (5.11)$$

The domain of values (x_N, N) for which (5.10) is valid constitutes a *zone of convergence*. When $V = 0$, (5.10) is the ordinary central limit approximation. For the value $V = V_0$ at which $\phi(iV)$ assumes its minimum, and at which $\mu_V = 0$, (5.10) reduces to (4.5b) when

$$|x^2/(2N\sigma_{V_0}^2)| \ll 1. \qquad (5.12)$$

III.5.2 The saddlepoint expansion

We have been considering the probability densities $f_0^{(N)}(x)$ for fixed x and varying N. Closely related to $f_0^{(N)}(x)$ is the probability density of the Nth mean increment $Y_N = N^{-1}\sum_1^N \xi_i$, given by $f_N^*(y) = Nf_0^{(N)}(Ny)$. It has been shown by Daniels that when y is fixed, $f_N^*(y)$ has an asymptotic series expansion in powers of N^{-1}, if the value $\mu_\theta = y$ is assumed by a conjugate mean within the convergence strip. A simple heuristic derivation of this series may be given along the following lines. By (4.3) $f_0^{(N)}(Ny) = \phi_0^N(i\theta)e^{\theta Ny} f_\theta^{(N)}(Ny)$ and $f_\theta^{(N)}(Ny)$ is the value of $f_\theta^{(N)}(x)$ at its mean. It is sufficient therefore to examine the asymptotic series for $f^{(N)}(0)$ associated with a density $f(x)$ with zero mean. Let us suppose that for N sufficiently large, (3.1) will be valid. We may then write

$$\sqrt{2\pi N\sigma^2}\, f^{(N)}(0) = (2\pi)^{-\frac{1}{2}} \int_{-\infty}^{\infty} \exp\{N \log\phi(z/\sqrt{N\sigma^2})\}\, dz, \qquad (5.13)$$

with $\phi(0) = 1$, $\phi'(0) = 0$. As $N \to \infty$, the exponent in the integrand is asymptotic to $-\frac{1}{2}z^2$, by L'Hôpital's rule. The function $\exp\{N\log\phi(z/\sqrt{N\sigma^2})\}$ may be written as $\exp\{-(z^2/2)\Phi(w)\}$ where $w = z/\sqrt{N}$ and $\Phi(w) = -2\log\phi(\sigma^{-1}w)/w^2$. Since $\Phi(w)$ is analytic in w at $w = 0$, and $\Phi(0) = 1$, a Taylor expansion of $\exp\{-(z^2/2)\Phi(w)\}$ in powers of w (z being regarded as a parameter) will have the form $\exp\{-(z^2/2)\}\cdot\{1 + E_1(z^2)w + E_2(z^2)w^2 \ldots\}$, where $E_n(z^2)$ is seen to be a polynomial in z^2 of degree n. If we now substitute in (5.13) and integrate formally term by term, odd powers of w drop out and we obtain the heuristic series

$$f^{(N)}(0) \sim (2\pi N\sigma^2)^{-\frac{1}{2}}[1 + \alpha_1 N^{-1} + \alpha_2 N^{-2} \ldots] \qquad (5.14)$$

where $\alpha_j = (2\pi)^{-\frac{1}{2}}\int_{-\infty}^{\infty}\exp\{-z^2/2\}E_{2j}(z^2)z^j\,dz$. This series is heuristic because the Taylor expansion in w only gives a valid representation of the integrand in its circle of convergence, i.e. for values $|w| = N^{-\frac{1}{2}}|z|$ less than R_Φ, the distance from $w = 0$ of the zero or singularity of $\phi(\sigma^{-1}w)$ nearest the origin. Even though the series (5.14) will not in general converge, it still has meaning as an *asymptotic series* (Copson, 1935).

A function $\psi(N)$ is said to have an asymptotic series expansion $\sum_0^\infty \psi_j N^{-j}$ if $\lim_{N\to\infty} N^K\{\psi(N) - \sum_0^K \psi_j N^{-j}\} = 0$ for every $K \geqslant 0$. Consequently, one has $\lim_{N\to\infty} \sum_0^K \psi_j N^{-j}/\psi(N) = 1$, i.e., every truncation of the series is an asymptotic representation of $\psi(N)$. The truncation to $K+1$ terms gives greater accuracy, in general, than the truncation to K terms. The asymptotic character of the series (5.14) may be understood from the preceding paragraph, for the domain of z values for which the Taylor expansion is valid is $|z| < N^{\frac{1}{2}}R_\Phi$, and this domain grows with N.

A formal derivation of (5.14) may be obtained as follows. In place of (5.13) we may write $f^{(N)}(0) = (2\pi)^{-1}\int_{-\infty}^{\infty}\exp\{N\log\phi(z)\}\,dz$. We may then introduce the transformation $\log\phi(z) = -\frac{1}{2}y^2$. The inverse function $z = Z(y)$ required for the transformation may be written as $Z(y) = \sum_1^\infty \alpha_n y^n$ with the aid of Lagrange's formula for the inversion of power series. The series (5.14) is then obtained from integration term by term, and its character as an asymptotic series† is a consequence of Watson's lemma (Copson, 1935). Discussions of saddlepoint integration may be found in Copson (1935), Jeffreys and Jeffreys (1962), and deBruijn (1958).

The first two terms of the saddlepoint expansion are found in this manner or from (5.14) to be

$$f^{(N)}(0) \sim \frac{1}{\sqrt{2\pi N\mu_2^2}}\left[1 + \frac{1}{N}\left\{\frac{3\mu_2\mu_4 - 9\mu_2^3 - 5\mu_3^2}{24\mu_2^3}\right\}\ldots\right] \qquad (5.15)$$

† Details are given in Note 4.

where μ_k is the kth moment of $f(x)$.

From (5.15) and (4.3), we have therefore for any suitable density $f(x)$, whether or not its first moment vanishes,

$$f^{(N)}(Ny) \sim \frac{\phi_0^N(i\theta)e^{\theta Ny}}{\sqrt{2\pi N\mu_{\theta 2}^{*2}}}\left[1 + \frac{1}{N}\left\{\frac{3\mu_{\theta 2}^*\mu_{\theta 4}^* - 9\mu_{\theta 2}^{*3} - 5\mu_{\theta 3}^{*2}}{24\mu_{\theta 2}^{*3}}\right\}\dots\right] \quad (5.16)$$

where $\mu_{\theta k}^*$ is the kth moment about the mean of the conjugate density $f_\theta(x)$ associated with the saddlepoint θ, i.e., $\mu_{\theta k}^* = \int (y - \mu_\theta)^k f_\theta(y)\,dy$. For computation purposes these moments may be obtained from the derivatives of the characteristic function at the saddlepoint. Explicit formulae are given by Daniels (1954). Other saddlepoint expansions appropriate to lattice processes and processes in continuous time may be found in Daniels, and also in Keilson (1963b). See Note 5.

III.6 Analogues for the lattice

(a) *The annulus of convergence*

In Section 6 of Chapter II simple properties of the generating function

$$\pi(u) = \sum_{-\infty}^{\infty} p_n u^n \quad (6.1)$$

of a lattice distribution were given. Corresponding to the convergence strip for the integral (1.1) defining the characteristic function, is the annulus of convergence for the series (6.1) defining the generating function. The series (6.1) can be decomposed as in Section 1, into non-positive and positive components, so that $\pi(u) = q\pi_I(u) + p\pi_{II}(u)$ where, for $0 < q < 1$,

$$\pi_I(u) = \sum_{-\infty}^{0} p_n u^n \Big/ \sum_{-\infty}^{0} p_n; \qquad \pi_{II}(u) = \sum_{1}^{\infty} p_n u^n \Big/ \sum_{1}^{\infty} p_n. \quad (6.2)$$

The series $\pi_I(u)$ will always be absolutely and uniformly convergent on and outside the unit circle, i.e., for $|u| \geqslant 1$; the series $\pi_{II}(u)$ will always be absolutely and uniformly convergent on and inside the unit circle, i.e., for $|u| \leqslant 1$. The two generating functions $\pi_I(u)$ and $\pi_{II}(u)$ are in consequence regular respectively in the exterior and interior of the unit circle. The series $\pi_I(u)$ may be convergent inside the unit circle. By virtue of its character as a power series in u^{-1}, its domain of convergence will be some annular region $\rho_{\min} < |u|$, and possibly the boundary $|u| = \rho_{\min}$, where $0 < \rho_{\min} \leqslant 1$. Similarly, the domain of convergence of $\pi_{II}(u)$ will be some annulus $|u| < \rho_{\max}$, and possibly the boundary $|u| = \rho_{\max}$, where $1 \leqslant \rho_{\max}$. The common annular region, called the *annulus of convergence*, always contains the unit circle. When $\rho_{\min} \neq \rho_{\max}$ the annulus has finite width and $\pi(u)$ is regular in its interior, $\rho_{\min} < |u| < \rho_{\max}$. A plot of $\pi(\rho)$ for real values of ρ in the annulus of convergence for two simple distributions is exhibited in Fig. III.4.

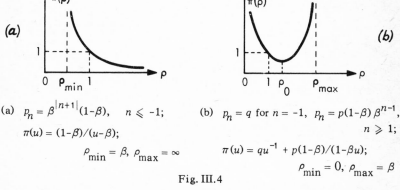

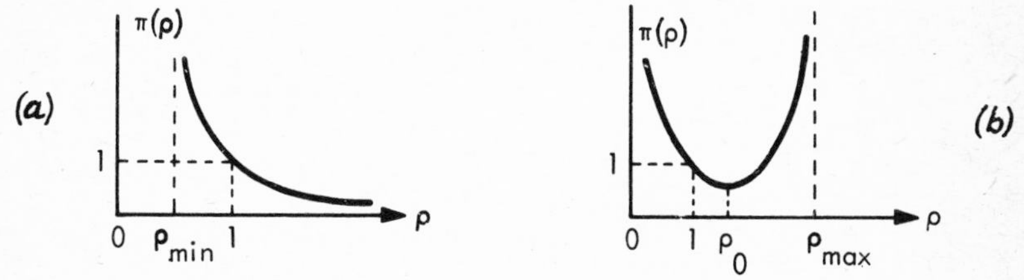

(a) $p_n = \beta^{|n+1|}(1-\beta), \quad n \leqslant -1;$
$\pi(u) = (1-\beta)/(u-\beta);$
$\rho_{\min} = \beta, \ \rho_{\max} = \infty$

(b) $p_n = q$ for $n = -1$, $p_n = p(1-\beta)\,\beta^{n-1}$, $n \geqslant 1;$
$\pi(u) = qu^{-1} + p(1-\beta)/(1-\beta u);$
$\rho_{\min} = 0, \ \rho_{\max} = \beta$

Fig. III.4

If $\pi(u)$ is analytic at $u = 1$ (this implies the existence of all moments of the lattice distribution), then $\pi(u)$ has a Laurent expansion about $u = 0$ in an annulus containing $u = 1$, and this must coincide with (6.1). The inversion formula of (II.6.13)

$$p_n = \frac{1}{2\pi i} \oint \pi(u) u^{-n-1}\, du$$

is then a direct consequence of Laurent's Theorem. The radii bounding an annulus of convergence of finite width are always associated with the presence of singularities of $\pi(u)$. As for the convergence strip, these singularities may be poles of arbitrary order, branch points, or essential singularities.

(b) *Completeness of the annulus of convergence*

Let $u = \rho e^{i\theta}$ where ρ and θ are the magnitude and argument of u. On the ray $\theta = 0$ corresponding to the positive real axis of u, the generating functions $\pi_I(\rho)$, $\pi_{II}(\rho)$, and $\pi(\rho)$ will be real, positive, and convex inside the annulus of convergence. The convexity follows from

$\pi''(\rho) = \sum_{-\infty}^{\infty} n(n-1)\, p_n \rho^{n-2} > 0$. The annulus of convergence of $\pi_{II}(u)$ will be said to be complete† if $\rho\,\pi_{II}'(\rho) \to +\infty$ as $\rho \to \rho_{\max}^{-}$. Similarly that for $\pi_I(u)$ will be said to be complete if $\rho\,\pi_I'(\rho) \to -\infty$ as $\rho \to \rho_{\min}^{+}$. When the lattice distribution has support on both the positive and non-positive sub-lattice, its generating function will be said to have a complete annulus of convergence if the annulus for both $\pi_I(u)$ and $\pi_{II}(u)$ are complete.

Consider for example the distribution on the positive lattice with $p_n = (a^n/n^2)/\sum_1^{\infty}(a^n/n^2)$ where $0 < a < 1$. Here $\rho_{\max} = a^{-1}$. $\pi(\rho) = \Sigma(a^n\rho^n/n^2)/\Sigma(a^n/n^2)$ has a finite limit as $\rho \to \rho_{\max}^{-}$, but $\rho\,\pi'(\rho)$ becomes infinite. The annulus of convergence is complete. For $p_n = (a^n/n^3)/\sum_1^{\infty}(a^n/n^3)$, however, $\pi(\rho)$ and $\rho\,\pi'(\rho)$ both have finite limits as $\rho \to \rho_{\max}^{-}$ and the annulus of convergence is not complete.

(c) *The conjugate family of distributions*

For lattice distributions, equation (2.1) when re-expressed in terms of the lattice probabilities becomes, for $e^{-V} = \rho$,

$$p_{\rho;n} = p_{1;n}\rho^n \Big/ \sum_{-\infty}^{\infty} p_{1;n}\rho^n . \tag{6.3}$$

Correspondingly,

$$\pi_\rho(u) = \pi_1(\rho u)/\pi_1(\rho). \tag{6.4}$$

For the set of values ρ in the interval $(\rho_{\min}, \rho_{\max})$, (6.3) will be a set of conjugate lattice probabilities. Equations (2.2) and (2.4) become

$$\mu_\rho = \pi_\rho'(1) = \rho \frac{d}{d\rho}\{\log \pi_1(\rho)\} \tag{6.5}$$

and

$$\sigma_\rho^2 = \rho \frac{d}{d\rho} \mu_\rho . \tag{6.6}$$

Theorems (2.1) and (2.2) were valid for any distributions. In particular, for lattice distributions† we have:

† Note that $\phi(U) = \Sigma p_n e^{inU} = \pi(e^{iU})$, and $\phi(iV) = \pi(e^{-V}) = \pi(\rho)$. Hence $\rho(d/d\rho) = -(d/dV)$. Completeness for the lattice is therefore the analogue of the more general completeness of Section 2.

Theorem III.6.1. For a non-degenerate conjugate family of lattice distributions, μ_ρ is a continuous monotonic increasing function of ρ. If the annulus of convergence is complete, for each value α in (x_L, x_M) there will be one member $\{p_{\rho;n}\}$ with $\mu_\rho = \alpha$. On the ray $\theta = 0$, inside the annulus of convergence $\pi(\rho)$ is a convex function of ρ. If the distributions have both positive and negative support and if this annulus is complete, $\pi(\rho)$ assumes a minimum inside the convergence strip.

(d) *The Central Limit Theorem and zones of convergence*

When a complete annulus of convergence is available, the reasoning of Sections 4 and 5 may be employed to make optimum use of the Central Limit Theorem. Suppose one is interested in the sum of N independent lattice random variables $X_N = \sum_1^N \xi_j$, for which $P\{\xi_j = n\} = p_{1;n}$ with generating function $\pi_1(u)$. The lattice distribution for X_N is the N-fold convolution $\{p_{1;n}^{(N)}\}$ (cf. Section 4 of Chapter II), with generating function $\pi_1^N(u)$. Let $\{p_{1;n}\}$ have mean μ_1 and variance σ_1^2. If $(n - N\mu_1)/(N^{2/3}\sigma_1)$ is large, direct application of the Central Limit Theorem is undesirable. Instead one works with the conjugate lattice distribution of parameter β for which $\mu_\beta = n/N$. Then, from (6.3),

$$p_{\beta;n}^{(N)} = \beta^n p_{1;n}^{(N)} \Big/ \pi_1^N(\beta). \tag{6.7}$$

We may now apply the Central Limit Theorem† to $p_{\beta;n}^{(N)}$ to obtain $p_{\beta;n}^{(N)} \sim (2\pi N \sigma_\beta^2)^{-\frac{1}{2}}$, provided the possible increments have for their difference

---

† The Central Limit Theorem takes the following form for the lattice.

Theorem III.6.2 (Gnedenko and Kolmogorov, 1954, p. 232). Let the independent, identically distributed random variables $\xi_1, \xi_2, \ldots$ take only integral values and have mean μ and finite variance $\sigma^2 \neq 0$. If the greatest common divisor of the differences of the values assumed by ξ_n with positive probability is one, then for $g_n^{(k)} = P\{\sum_1^k \xi_i = n\}$, we have

$$g_n^{(k)} \sim \frac{e^{-(n-k\mu)^2/2k\sigma^2}}{\sqrt{2\pi k \sigma^2}} \qquad \text{as } k \to \infty.$$

Note that when the only increments assumed by a lattice process N_k are +1 and −1, the greatest common divisor is 2 and the theorem is not applicable. This corresponds to the fact that $g_n^{(k)} = 0$ for even n when k is odd, and for odd n when k is even. For the sum of two successive increments, the values 0 and ±2 are assumed, and the process $N_k' = \frac{1}{2} \sum_1^{2k} \xi_i$ has increments 0 and ±1. The theorem is therefore applicable to N_k'.

a greatest common divisor of one. Hence from (6.7) the "saddlepoint" approximation becomes the asymptotic representation, as $N \to \infty$, †

$$p_{1;n}^{(N)} \sim \pi_1^N(\beta)\, \beta^{-n}(2\pi N \sigma_\beta^2)^{-\frac{1}{2}} . \tag{6.8}$$

When n is fixed, and $N \to \infty$, $\mu_\beta \to 0$ and $\beta \to \rho_0$, if there is a value ρ_0 in the strip for which $\pi_1(\rho)$ takes on its minimum value. We have then a second asymptotic representation (c.f. equation (4.5b))

$$p_{1;n}^{(N)} \sim \pi_1^N(\rho_0)\, \rho_0^{-n}\, (2\pi N \sigma_{\rho_0}^2)^{-\frac{1}{2}} . \tag{6.9}$$

The approximation contained in (6.9) is weaker but structurally simpler than that contained in (6.8).

III.7 Modification for processes in continuous time

In place of the partial sum $\sum_1^N \xi_i$ for processes in discrete time, one usually deals in continuous time with

$$X(t) = \sum_1^{K(t)} \xi_i + vt + X_D(t) \tag{7.1}$$

where $K(t)$ is an auxiliary Poisson process and $X_D(t)$ is a Wiener-Lévy process (II.3d). Additive homogeneous processes with this characterization were discussed in Chapter II. There we saw in Section II.7 that for ξ_i having distribution $A(x)$ and characteristic function $\alpha(z)$, and for Poisson increments of mean arrival rate ν, the distribution function $G(x,t)$ for the process $X(t)$ has the characteristic function

$$\gamma(z,t) = e^{-\nu t[1-\alpha(z)]}\, e^{-Dz^2 t}\, e^{izvt} . \tag{7.2}$$

The function $\gamma(z,\nu^{-1})$ is the characteristic function for the distribution $G(x,\nu^{-1})$. Since

$$G(x,t) = G^{(\nu t)}(x,\nu^{-1}), \tag{7.3}$$

the methods and results of Sections 4 and 5 are applicable when νt is a large integer, with $F(x)$ replaced by $G(x,\nu^{-1})$. The continuous time case has a few special characteristics worth recording.

---

† Correction terms are given by Daniels (1954) and Good (1957).

Theorem III.7.1. If for the homogeneous process (7.1), the increment distribution $A(x)$ has a complete characteristic function, then the c.f. $\gamma(z,t)$ is also complete for all time t. If, further, the increments ξ_i are non-negative (non-positive), the family of conjugate distributions $G_V(x,t)$ will have associated means $\mu_V(t)$ taking on all values in the interval $[vt,\infty)$ (or $(-\infty,vt]$) when $D = 0$, and $(-\infty,\infty)$ when $D \neq 0$. If the increments ξ_i have both negative and positive support, the means $\mu_V(t)$ assume all real values for $t > 0$.

The completeness is straightforward, and follows from the behaviour of the derivative of $\gamma(z,t)$ at the boundaries of the convergence strip of $\alpha(z)$. The range of values $\mu_V(t)$ may be inferred from Theorem 2.2. Henceforth completeness will be assumed.

The accommodation of the convergence strip to the range of values assumed by the process is striking. In discrete time X_N could assume any value in $[Nx_L, Nx_M]$, and for each such value one conjugate distribution was available having that mean. In continuous time, the same accommodation has been placed in evidence.

Theorem III.7.2. Consider the homogeneous process $X(t)$ of (7.1) governed by increment distribution $A_0(x)$, frequency ν_0, diffusion coefficient D_0, and drift rate v_0. Let $A_0(x)$ have the conjugate family of distributions $A_V(x)$. Then the distribution $G_V(x,t)$ conjugate to $G_0(x,t)$ is the Green distribution for another process $X_V(t)$ of type (7.1) governed by increment distribution $A_V(x)$, and parameters

$$\text{(a)}\quad \nu_V = \nu_0\,\alpha_0(iV); \qquad \text{(b)}\quad D_V = D_0; \qquad \text{(c)}\quad v_V = v_0 - 2D_0 V. \tag{7.4}$$

Equation (7.4) follows from (7.1), (2.1a) and simple algebra. From (7.4) the mean and variance of $G_V(x,t)$ are

$$\mu_V(t) = \nu_V t\,\mu_V + v_V t = \nu_0 t \int_{-\infty}^{\infty} x\,e^{-Vx}\,dA_0(x) + (v_0 - 2D_0 V)t \tag{7.5}$$

and

$$\sigma_V^2(t) = \nu_0 t \int_{-\infty}^{+\infty} x^2\,e^{-Vx}\,dA_0(x) + 2D_0 t\,. \tag{7.6}$$

The integral form of the Central Limit Theorem for continuous time, valid for any of the conjugate processes $X_V(t)$, follows from (7.3) and states that, for x fixed as $t \to \infty$,

$$G_V(\sigma_V(t)x + \mu_V(t), t) \sim \frac{1}{\sqrt{2\pi}} \int_{-\infty}^{x} e^{-y^2/2} dy . \tag{7.7}$$

Suppose one wished to estimate $G(x,t)$ for arbitrary fixed x and large t. The saddlepoint approximation implicit in (4.7) is obtained from (7.7) and the optimum conjugate transformation of Section 4. The saddlepoint is determined by equation (4.1), which now has the form

$$\mu_\theta(t) = x \tag{7.8}$$

or from (7.5),

$$(v_0 - 2D_0\theta) + \nu_0 \int_{-\infty}^{\infty} x\, e^{-\theta x} dA_0(x) = x/t . \tag{7.9}$$

Since for any V in the convergence strip

$$G_0(x,t) = \gamma_0(iV,t) \int_{-\infty}^{x} e^{Vx'} dG_V(x', t) , \tag{7.10a}$$

we have for $V = \theta$,

$$G_0(x,t) = \gamma_0(i\theta,t) \int_{-\infty}^{x} e^{\theta x'} dG_\theta(x', t) \tag{7.10b}$$

where θ is found from the saddlepoint equation (7.9). The Central Limit Theorem may now be applied to $G_\theta(x,t)$ in (7.10). For arbitrary distribution $A(x) = A_0(x)$ and corresponding $G_0(x,t)$, a result comparable to (4.7) may be written down. Of greater interest is the analogue of (4.5), when the local density $g_0(x,t)$ is smooth enough†. From (7.10b) we have‡ $g_0(x,t) \sim e^{\theta x} \gamma_0(i\theta,t)\{2\pi\sigma_\theta^2(t)\}^{-\frac{1}{2}}$ as $t \to \infty$. Hence from (7.2) and (7.6), we have the saddlepoint approximation for continuous time

$$g_0(x,t) \sim \frac{\exp[\theta x - \nu_0 t\{1 - \alpha_0(i\theta)\} + D_0\theta^2 t - v_0\theta t}{\sqrt{2\pi\{2D_0 t + \nu_0 t \int x^2 e^{-\theta x} dA_0(x)\}}} . \tag{7.11}$$

If x is fixed and $t \to \infty$, equation (4.1) implies that the saddlepoint $i\theta$ approaches the point iV_0 at which $\gamma(iV, \nu_0^{-1})$ takes on its minimum‡,

† A more precise result is given in Note 6.

‡ If $D_0 = 0$, $v_0 = 0$, and $A_0(x)$ has purely positive support, the minimum needed for the asymptotic character of (7.11) is unavailable. Compare the discussion below equation (4.5a).

and one obtains the analogue of (4.5b) by the substitution $\theta = V_0$ in (7.11).

When $G_0(x,t)$ is differentiable at x, we obtain from (7.7) and (7.10a) the set of Gaussian approximations

$$g_0(x,t) \approx e^{Vx}\gamma_0(iV,t)\,\frac{\exp[-\{x-\mu_V(t)\}^2/2\sigma_V^2(t)]}{\sqrt{2\pi\sigma_V^2(t)}} \tag{7.12}$$

the analogue of (5.10). The zone of convergence in which (7.12) will be valid may be inferred from (5.11). For simplicity suppose that $D = 0$, and that the departure of the increment distribution $A_0(x)$ from normality is not great (cf. Section 5.1). For $t = \nu_V^{-1}$, $G_V(x, \nu_V^{-1})$ will itself not be far from normality. We may then apply (5.2) with $N = t/\nu_V^{-1} = \nu_V t$, and $f_0(x) = g_V(x, \nu_V^{-1})$ to obtain the conditions

$$\text{(a)}\ \nu_V t \gg 1; \qquad \text{(b)}\ \frac{|x - \mu_V(t)|}{(\nu_V t)^{2/3}\sigma_V(\nu_V^{-1})} \ll 1 \tag{7.13}$$

where, from (7.6), $\sigma_V(\nu_V^{-1}) = \{\int x^2\, dA_V(x)\}^{\frac{1}{2}}$. Equation (7.12) with $V = 0$ is the diffusion approximation.

III.8 Lattice processes in continuous time

For the lattice process $N_1(t) = \sum_1^{K(t)} \xi_k$ in continuous time, the increments ξ_k are the lattice random variables of Section 6 with $P\{\xi_j = n\} = e_{1;n}$ and generating function $\epsilon_1(u)$. The process $N_1(t)$ then has state probabilities $g_{1;n}(t)$ given by equation (II.4.12) with the generating function

$$\gamma_1(u,t) = \exp[-\nu_1 t\{1 - \epsilon_1(u)\}]. \tag{8.1}$$

Associated with $\{e_{1;n}\}$ will be the conjugate distributions described in Section 6 with increment probabilities $\{e_{\rho;n}\}$ and generating function $\epsilon_\rho(u) = \epsilon_1(\rho u)/\epsilon_1(\rho)$. The generating function conjugate to (8.1) is, by (6.4),

$$\gamma_\rho(u,t) = \exp[-\nu_\rho t\{1 - \epsilon_\rho(u)\}] \tag{8.2}$$

where $\nu_\rho = \nu_1\,\epsilon_1(\rho)$, and (8.2) is the generating function for the conjugate process $N_\rho(t)$ with frequency ν_ρ and increment probabilities $\{e_{\rho;n}\}$. If the annulus of convergence of $\epsilon_1(u)$ is complete, this completeness will carry over to $\gamma_1(u,t)$. From (6.5) and (6.6) the mean and variance of $N_\rho(t)$ are

$$\mu_\rho(t) \;=\; \nu_\rho t \mu_\rho \;=\; \nu_1 t \sum_{-\infty}^{\infty} \rho^n n e_{1;n} \qquad (8.3)$$

and

$$\sigma_\rho^2(t) \;=\; \nu_\rho t (\mu_\rho^2 + \sigma_\rho^2) \;=\; \nu_1 t \sum_{-\infty}^{\infty} \rho^n n^2 e_{1;n} \,. \qquad (8.4)$$

From (6.3), for any ρ in the annulus of convergence, and in particular for $\rho = \beta$, the saddlepoint,

$$g_{1;n}(t) \;=\; \rho^{-n} \gamma_1(\rho,t) g_{\rho;n}(t) \;=\; \beta^{-n} \gamma_1(\beta,t) g_{\beta;n}(t). \qquad (8.5)$$

For given n and t, the saddlepoint β is determined by the saddlepoint equation $\mu_\beta(t) = n$, i.e.,

$$\nu_\beta \mu_\beta \;=\; \frac{n}{t}. \qquad (8.6)$$

When increments are non-negative, completeness assures that this equation will have a unique positive solution β for every non-negative integer n. When both positive and negative values are assumed, completeness assures a solution for every integer n. The Central Limit Theorem for the lattice applied to $g_{\beta;n}(t)$ gives $g_{\beta;n}(t) \approx [2\pi\sigma_\beta^2(t)]^{-\frac{1}{2}}$. From (8.5) we then have the saddlepoint approximation, valid when $\nu_\beta t \gg 1$,

$$g_{1;n}(t) \;\approx\; \frac{\beta^{-n} \exp[-\nu_1 t\{1 - \epsilon_1(\beta)\}]}{\sqrt{2\pi \nu_\beta t (\sigma_\beta^2 + \mu_\beta^2)}} \,. \qquad (8.7)$$

The Gaussian approximation obtained from (8.7) when one is interested in a succession of values n_t in the zone of convergence about $\mu_\beta(t)$, is

$$g_{1;n}(t) \;\approx\; \frac{\beta^{-n} \exp[-\nu_1 t\{1 - \epsilon_1(\beta)\}]}{\sqrt{2\pi \sigma_\beta^2(t)}} \exp[-\{n - \mu_\beta(t)\}^2/\{2\sigma_\beta^2(t)\}] \qquad (8.8)$$

and (8.8) will be accurate when

$$\frac{|n - \nu_\beta \mu_\beta t|}{(\nu_\beta t)^{2/3}\sqrt{\mu_\beta^2 + \sigma_\beta^2}} \ll 1, \quad \text{and} \quad \nu_\beta t \gg 1. \qquad (8.9)$$

The conditions stated for the accuracy of (8.7) and (8.8) are again based on the assumption that the increment distribution $\{e_{1;n}\}$ is not too far from normality. The two zones of greatest interest are associated with the value $\beta = 1$, for the central limit approximation, for which

$$g_{1;n}(t) \approx \{2\pi\sigma_1^2(t)\}^{-\frac{1}{2}} \exp[-\{n - \mu_1(t)\}^2/2\sigma_1^2(t)]; \qquad (8.10)$$

and $\beta = \rho_0$, the value in the annulus of convergence at which $\epsilon_1(\rho)$ assumes its minimum, for which the zone of convergence (8.9) corresponds to $\mu_\beta = 0$, i.e., to values of n fixed in time. We then have

$$g_{1;n}(t) \approx \rho_0^{-n}\{2\pi\sigma_{\rho_0}^2(t)\}^{-\frac{1}{2}} \exp[-\nu_1 t + \nu_1 \epsilon_1(\rho_0)t - n^2\{2\sigma_{\rho_0}^2(t)\}^{-1}].$$

(8.11)

The approximation sign in (8.7) may be replaced by the asymptotic sign $\sim$ and the stipulation that $t \to \infty$, by virtue of the Central Limit Theorem for the lattice. Equation (II.9.7) then follows from this asymptotic relation and (II.9.1).

CHAPTER IV

THE PASSAGE PROBLEM FOR HOMOGENEOUS SKIP-FREE PROCESSES

The present chapter has as its principal objective a demonstration of the occurrence of the Green's function as an entity in the passage-time densities associated with a certain class of spatially homogeneous "skip-free" processes. For a lattice process $N(t)$ in continuous time with this skip-free property, one finds that the density $s_{n_0}(\tau)$ for the first passage time τ from the state $n_0 > 0$ to the state $n = 0$ is related to the Green's function $g_n(t)$ of the homogeneous process by

$$s_{n_0}(\tau) = \frac{n_0}{\tau} g_{-n_0}(\tau). \tag{0.1}$$

A relation of equal simplicity is available for similar processes in discrete time on the lattice and for a broad class of skip-free processes on the continuum. The central limit methods of Chapter III may be invoked to exhibit from such relations the asymptotic behaviour of the passage-time density. The diffusion approximation to the passage-time density, and the tail of the passage-time density correspond to two distinct zones of convergence.

IV.1 Skip-free processes

A random variable of basic interest in probability theory is the *passage time*. For a Markov process with a given single-step transition distribution $A(x',x)$, and initial value x_0, at $t = 0$, one is interested in the time κ (or the time τ for processes in continuous time) at which the process first reaches any state of a prescribed interval or set of states. One wishes to find the probability $s^{(\kappa)}$ or probability density $s(\tau)$ for this passage time.

For the spatially homogeneous processes, applications of interest deal typically with (A) the passage time from some value $x_0 > 0$ to the interval $(-\infty,0)$, or (B) the passage time from x_0 on a finite interval $(0,L)$ to the states outside this interval. For problems of the first type, a simple relationship may often be exhibited between the passage-time distribution and the Green distribution of the spatially homogeneous process. This relationship enables one to apply the Central Limit methods developed in Chapter III and obtain useful asymptotic approximations.

In this chapter we will develop such relations for a family of homogeneous processes having the following *skip-free* property in the direction of the absorbing region:

A process will be said to be skip-free negative (positive) if a sample cannot pass from x_2 to $x_1 < x_2$ ($> x_2$) without passing through all intervening states. Examples of the skip-free negative processes are:

(1) The homogeneous diffusion process (Section II.3). This process is also skip-free positive.

(2) The drift processes (Section II.3) with negative drift velocity v and positive increments.

(3) Any sum of processes of type 1 and type 2.

(4) The lattice processes where only unit increments are permitted in the negative direction. An important sub-case is the lattice process with Bernoulli increments (Section II.8), which is skip-free in both directions.

IV.2 The image method for lattice walks with Bernoulli increments

A type of mirror-like symmetry present in many problems of applied mathematics permits one to employ the very effective classical method of images. The same method provides a key to the discussion of the homogeneous lattice processes with Bernoulli increments, modified by the presence of absorbing or reflecting boundaries. When the symmetry required is not present, it may be introduced by a suitable conjugate transformation of the type discussed in Chapter III. All of the results of this section and some of the methods are well known, and are either to be found in Feller (1957) or other standard sources. Our purpose is to demonstrate one's ability to discuss the discrete and continuous time walks with Bernoulli increments without recourse to generating functions or Laplace transformation, and to provide a background for the subsequent generalizations. The methods employed impart to the results a simple probabilistic structure. We first consider the passage problem for one absorbing boundary.

(A) *One absorbing boundary*

Consider the symmetric homogeneous lattice process N_k with e_{-1}

$= e_{+1} = \frac{1}{2}$ commencing at $n_0 > 0$. The notation employed is that of Section II.4. Because of the symmetry of e_n about $n = 0$, the Green probabilities $g_n^{(k)} = P\{N_k = n | N_0 = 0\}$ are also symmetric about $n = 0$, i.e.,

$$g_n^{(k)} = g_{-n}^{(k)}. \tag{2.1}$$

If one were interested in the time κ at which N_k first reaches the state 0, one could take advantage of the symmetry of (2.1) to obtain the distribution of κ in the following manner. Let $q_n^{(k)}$ be the joint probability that $N_k = n$ and that the state $n = 0$ has not been reached at time k. Then

$$q_0^{(k)} = 0 \tag{2.2}$$

and

$$q_n^{(k)} = \frac{1}{2} q_{n-1}^{(k-1)} + \frac{1}{2} q_{n+1}^{(k-1)} \quad \text{for } k \geqslant 1,\ n \geqslant 1. \tag{2.3}$$

The initial condition is

$$q_n^{(0)} = \delta_{n,n_0}, \quad \text{for all } n. \tag{2.4}$$

The Green probabilities $\{g_n^{(k)}\}$ satisfy (2.3) for all n. Hence the set of probabilities $\{g_{n+n_0}^{(k)}\}$ and $\{g_{n-n_0}^{(k)}\}$ both satisfy (2.3) for all n. The symmetry of (2.1) may be exploited to satisfy (2.2) and (2.3) with the solution

$$q_n^{(k)} = g_{n-n_0}^{(k)} - g_{n+n_0}^{(k)} \tag{2.5}$$

for which $q_n^{(0)} = \delta_{n,n_0} - \delta_{n,-n_0}$. It may now be observed that the system of equations (2.2), (2.3) and (2.4) involves only positive values of n, i.e., these equations uniquely determine the set of probabilities $q_n^{(k)}$, since these values are generated recursively by (2.3). Since (2.5) satisfies (2.4) for all $n \geqslant 1$, the positive components of (2.5) are the unique solution of the equation system. Equation (2.5) may be interpreted as resulting from an initial distribution consisting of the actual concentration at n_0 and a fictitious negative "image" at $-n_0$. Hence the name *method of images*.

The Green probabilities for the symmetric lattice walk with Bernoulli increments have already been exhibited in Section II.8. They are given by $g_n^{(k)} = \beta_n^{(k)}$ where

$$\beta_n^{(k)} = \begin{cases} 2^{-k}\binom{k}{\frac{1}{2}(k-n)}, & k-n \text{ even} \\ 0 \quad , & k-n \text{ odd.} \end{cases} \tag{2.6}$$

If we denote the probability that $\kappa = k$ when $N_0 = n_0$ by $s_{n_0}^{(k)}$, we have from $s_{n_0}^{(k)} = \frac{1}{2}q_1^{(k-1)}$ and (2.5),

$$s_{n_0}^{(k)} = \frac{1}{2}\left\{\beta_{1-n_0}^{(k-1)} - \beta_{1+n_0}^{(k-1)}\right\} = \frac{n_0}{k}\beta_{n_0}^{(k)} = \frac{n_0}{k}g_{-n_0}^{(k)}. \tag{2.7}$$

In (2.7) we have made use of the identities $\beta_n^{(k)} = \beta_{-n}^{(k)}$, and

$$\beta_{n-1}^{(k)} - \beta_{n+1}^{(k)} = 2n(k+1)^{-1}\beta_n^{(k+1)}. \tag{2.7$'$}$$

The latter identity is a subcase of a more general relation (3.2) derived in Section 3.

When $e_1 = p$, $e_{-1} = q$ and $p \neq q$, (2.3) must be replaced by

$$q_n^{(k)} = p\,q_{n-1}^{(k-1)} + q\,q_{n+1}^{(k-1)}. \tag{2.8}$$

The lack of symmetry for $p \neq q$ is only superficial. If the conjugate transformation

$$q_n^{(k)} = (4pq)^{k/2}\,(p/q)^{\frac{1}{2}(n-n_0)}\,r_n^{(k)} \tag{2.9}$$

is introduced one finds that $r_n^{(k)}$ satisfies equations (2.2), (2.3) and (2.4). Since we have seen that this system has a unique solution we infer that $r_n^{(k)} = \beta_{n-n_0}^{(k)} - \beta_{n+n_0}^{(k)}$. We then have from (2.9)

$$q_n^{(k)} = (4pq)^{k/2}(p/q)^{\frac{1}{2}(n-n_0)}\{\beta_{n-n_0}^{(k)} - \beta_{n+n_0}^{(k)}\} = g_{n-n_0}^{(k)} - \left(\frac{q}{p}\right)^{n_0} g_{n+n_0}^{(k)} \tag{2.10}$$

where

$$g_n^{(k)} = (p/q)^{\frac{1}{2}n}\,(4pq)^{\frac{1}{2}k}\,\beta_n^{(k)} \tag{2.10$'$}$$

obtained from (II.8.7) is the Green probability for the unsymmetric process. The passage-time probability is again given from (2.10) and (2.7′) by

$$s_{n_0}^{(k)} = q\,q_1^{(k-1)} = \frac{n_0}{k} g_{-n_0}^{(k)}. \tag{2.11}$$

For the associated processes $N(t)$ in continuous time (Section II.2), when transition epochs occur at frequency ν, the corresponding results are immediately available. The probability density for the time of the kth transition is $\nu(\nu t)^{k-1}\,e^{-\nu t}/(k-1)!$. If we denote the passage time at which $n = 0$ is first reached from the state n_0 by τ and its probability density by $s_{n_0}(\tau)$, we have clearly

$$s_{n_0}(\tau) = \sum_{k=1}^{\infty} \{\nu(\nu\tau)^{k-1}\, e^{-\nu\tau}/(k-1)!\}\, s_{n_0}^{(k)}. \tag{2.12}$$

Hence from (2.12) we have $s_{n_0}(\tau) = n_0\,\tau^{-1} \sum_0^{\infty} \{(\nu\tau)^k e^{-\nu\tau}/k!\}\, g_{-n_0}^{(k)}$. But from (II.4.12), the series is the Green probability $g_{-n_0}(\tau)$ for the associated process in continuous time, and (2.13) becomes the simple result

$$s_{n_0}(\tau) = \frac{n_0}{\tau} g_{-n_0}(\tau). \tag{2.13}$$

In Section II.8 we saw that the Green probability for continuous time for Bernoulli increments is

$$g_n(t) = \left(\frac{p}{q}\right)^{n/2} e^{-\nu t}\, I_n(\nu t\sqrt{4pq}) \tag{2.14}$$

where I_n is the modified Bessel function of order n. The relation (2.13) with $g_n(t)$ given by (2.14) is a classical result.

In the same way we can exhibit $q_n(t)$, the joint probability that $N(t) = n$ and that the state $n = 0$ has not been reached. For $q_n(t) = \sum_0^{\infty} \{(\nu t)^k e^{-\nu t}/k!\}\, q_n^{(k)}$ and from (2.10) we have

$$q_n(t) = g_{n-n_0}(t) - \left(\frac{q}{p}\right)^{n_0} g_{n+n_0}(t). \tag{2.15}$$

(B) *The reflecting boundary*

The homogeneous Bernoulli walk of part A modified by a reflecting boundary at $n = 0$ is often of interest. At a transition epoch, there is a probability p of transition from the boundary state $n = 0$ to $n = 1$, and probability q of remaining at zero. The new process which we designate by N'_k is thereby confined to the non-negative integers. Let $p_n^{(k)} = P\{N'_k = n | N'_0 = n_0\}$. Then equation (2.8) is replaced by

$$p_n^{(k)} = p\,p_{n-1}^{(k-1)} + q\,p_{n+1}^{(k-1)}, \qquad k \geqslant 1,\ n \geqslant 1, \tag{2.16}$$

and

$$p_0^{(k)} = q(p_1^{(k-1)} + p_0^{(k-1)}), \qquad k \geqslant 1, \tag{2.17}$$

and the initial condition is given by $p_n^{(0)} = \delta_{n,n_0}$.

When $p = q = \frac{1}{2}$ we may take advantage of the overt symmetry of $g_n^{(k)}$ about $n = 0$, to exhibit the state probabilities in compact form. We note that (2.17) may be inferred from (2.16) and from the equation $\sum_{n=0}^{\infty} p_n^{(k)} = 1$ describing the conservation of probability. We also observe that for the corresponding homogeneous symmetric process N_k^* with $p = q = \frac{1}{2}$ and $p_n^{*(0)} = \frac{1}{2}\delta_{n,n_0} + \frac{1}{2}\delta_{n,-n_0-1}$, the equations of continuity coincide with (2.16) for $n \geqslant 1$, and by virtue of the symmetry, $\sum_{n=0}^{\infty} p_n^{*(k)} = \frac{1}{2}$ for all k. Hence we infer that $p_n^{(k)} = 2p_n^{*(k)}$, i.e., that for $p = q = \frac{1}{2}$,

$$p_n^{(k)} = g_{n-n_0}^{(k)} + g_{n+n_0+1}^{(k)} = \beta_{n-n_0}^{(k)} + \beta_{n+n_0+1}^{(k)}, \qquad n \geqslant 0,\ k \geqslant 0, \tag{2.18}$$

where $g_n^{(k)} = \beta_n^{(k)}$ is the Green's function for the symmetric walk given by (2.6). For the symmetric process in continuous time, we have, as for (2.15),

$$p_n(t) = g_{n-n_0}(t) + g_{n+n_0+1}(t), \qquad n \geqslant 0, \quad t \geqslant 0, \tag{2.19}$$

where $g_n(t)$ is the Green probability $e^{-\nu t} I_n(\nu t)$ of (2.14).

When the walk is not symmetric, i.e., when $p \neq q$, the state probabilities $p_n^{(k)}$ and $p_n(t)$ have a slightly more complex form. Simplicity of form then resides in a set of flow quantities basic to the system motion. Let the net flow of probability from $(n-1)$ to n at the $(k+1)$st step be denoted by

$$j_n^{(k)} = p\,p_{n-1}^{(k)} - q\,p_n^{(k)}, \qquad n \geqslant 1, \quad k \geqslant 0. \tag{2.20}$$

These flow quantities may be written down directly from the results for the passage problem. From (2.16_n), (2.16_{n-1}) and (2.20), we have

$$j_n^{(k)} = p\, j_{n-1}^{(k-1)} + q\, j_{n+1}^{(k-1)}, \qquad n \geqslant 2, \quad k \geqslant 1, \qquad (2.21_a)$$

and

$$j_1^{(k)} = q\, j_2^{(k-1)}, \qquad k \geqslant 1. \qquad (2.21_b)$$

The set of equations (2.21) are identical with those of sub-section A governing the passage problem, the solution of which, when $N_0 = n_0$, is (2.10). Since $j_n^{(0)} = p\,\delta_{n,n_0+1} - q\,\delta_{n,n_0}$ from (2.20), the flow quantities are simply related to the Green probabilities by

$$j_n^{(k)} = p\left\{g_{n-n_0-1}^{(k)} - \left(\frac{q}{p}\right)^{n_0+1} g_{n+n_0+1}^{(k)}\right\} - q\left\{g_{n-n_0}^{(k)} - \left(\frac{q}{p}\right)^{n_0} g_{n+n_0}^{(k)}\right\}. \qquad (2.22)$$

(From $j_n^{(k)}$ we may obtain the state probabilities. For we have, for $n \geqslant 1$, $p_n^{(k+1)} - p_n^{(k)} = p\,p_{n-1}^{(k)} + q\,p_{n+1}^{(k)} - (p+q)p_n^{(k)}$, and from (2.20) $p_n^{(k+1)} - p_n^{(k)} = j_n^{(k)} - j_{n+1}^{(k)}$. Similarly $p_0^{(k+1)} - p_0^{(k)} = -j_1^{(k)}$, and the state probabilities are obtained by summation. The result is not very useful.)

For the processes in continuous time, the net rate at which probability flows from $(n-1)$ to n at time t is

$$j_n(t) = \nu p\, p_{n-1}(t) - \nu q\, p_n(t), \qquad n \geqslant 1, \quad t > 0. \qquad (2.23)$$

Since $j_n(t) = \nu \sum_0^\infty \{(\nu t)^k\, e^{-\nu t}/k!\}\, j_n^{(k)}$, we have from (2.22) and (II.4.12)

$$j_n(t) = \nu p\,\{g_{n-n_0-1}(t) - \left(\frac{q}{p}\right)^{n_0+1} g_{n+n_0+1}(t)\}$$
$$- \nu q\,\{g_{n-n_0}(t) - \left(\frac{q}{p}\right)^{n_0} g_{n+n_0}(t)\}. \qquad (2.24)$$

The rate of change of the state probabilities obtained from the continuity equations (II.4.5) is

$$\frac{d}{dt} p_n(t) = j_n(t) - j_{n+1}(t) \qquad (2.24')$$

valid for all n if we define $j_0(t)$ to be zero. Thus $p_n(t)$ may be exhibited from (2.24) as the sum of four indefinite integrals of modified Bessel functions, a familiar result.† For study of the asymptotic rate of approach of $p_n(t)$ to its ergodic distribution, when the process is ergodic, (2.24′) and (2.24) are of direct interest.

(C) *Two absorbing boundaries*

Instead of considering passage of the homogeneous process N_k to the single state $n = 0$, one might consider passage to either the state $n = 0$ or to the state $n = R$ where $0 < n_0 < R$. Let $q_n^{(k)}$ be the probability that $N_k = n$ and that neither $n = 0$ nor $n = R$ have been reached. Then $q_0^{(k)} = q_R^{(k)} = 0$ and

$$q_n^{(k)} = p\,q_{n-1}^{(k-1)} + q\,q_{n+1}^{(k+1)} \quad \text{for } k \geqslant 1, \quad 1 \leqslant n \leqslant R-1. \tag{2.25}$$

When $p = q = \frac{1}{2}$, the solution is given in terms of a set of negative images at $n = -n_0 + 2Rj$ for $j = 0, \pm 1, \pm 2$, etc., and a set of positive images at $n = n_0 + 2Rj$ for $j = \pm 1, \pm 2$, etc. The symmetric solution is then

$$q_n^{(k)} = \sum_{j=-\infty}^{\infty} \left\{ \beta_{n-n_0+2Rj}^{(k)} - \beta_{n+n_0+2Rj}^{(k)} \right\}. \tag{2.26}$$

When $p \neq q$, the symmetrizing transformation (2.9) may again be employed, and one obtains

$$q_n^{(k)} = (4pq)^{k/2}\,(p/q)^{\frac{1}{2}(n-n_0)} \sum_{-\infty}^{\infty} \left\{ \beta_{n-n_0+2Rj}^{(k)} - \beta_{n+n_0+2Rj}^{(k)} \right\}. \tag{2.27}$$

The infinite series may be summed by the following generating-function method. Consider the generating function $g(w) = \sum_{-\infty}^{\infty} g_n w^n$. Let the $2R$th roots of unity be denoted by

$$\chi_\alpha = e^{i\pi\alpha/R}, \qquad \alpha = 1,2,\ldots,2R. \tag{2.28}$$

One may employ the identity

---

† When $n_0 = 0$, we have from (2.24) and (3.2′) the simpler result $j_n(t) = \frac{n}{t} g_n(t)$, and $p_n(t) = \delta_{n,0} + \int_0^t s^{-1}\{n\,g_n(s) - (n+1)\,g_{n+1}(s)\}\,ds$. This result has been obtained by K. Stange (1964, *Unternehmenforsch.*, *8*, 1-24) by inversion of a Laplace transform-generating function. Numerical results are plotted there for this case.

$$\sum_{\alpha=1}^{2R} \chi_\alpha^n = 2R \quad \text{if} \quad n = 0, \pm 2R, \pm 4R, \text{etc.}, \tag{2.29}$$

$$= 0 \quad \text{otherwise,}$$

to obtain

$$\sum_{j=-\infty}^{+\infty} g_{m+2Rj} = \frac{1}{2R} \sum_{\alpha=1}^{2R} \chi_\alpha^{-m}\, g(\chi_\alpha), \tag{2.30}$$

and thereby convert the infinite series (2.27) into a finite sum. Since the generating function of $\beta_n^{(k)}$ is $2^{-k}(u+u^{-1})^k$, $g(\chi_\alpha) = \cos^k(\pi\alpha/R)$. From (2.30) we then have

$$\sum_{j=-\infty}^{\infty} \beta_{m+2Rj}^{(k)} = \frac{1}{R} \sum_{\alpha=1}^{R} \cos(m\pi\alpha/R)\cos^k(\pi\alpha/R) \tag{2.31}$$

and (2.27) becomes

$$q_n^{(k)} = (4pq)^{k/2}(p/q)^{\frac{1}{2}(n-n_0)}\left\{\frac{2}{R}\sum_{\alpha=1}^{R-1} \sin(n\pi\alpha/R)\sin(n_0\pi\alpha/R)\cos^k(\pi\alpha/R)\right\}. \tag{2.32}$$

The passage-time probabilities are obtained from (2.32). If we denote by $A_{n_0}^{(k)}$ the probability that the process leaves the interval set $\{n = 1, 2, \ldots, R-1\}$ for the first time via $n = 0$ at the kth step, and that via $n = R$ by $B_{n_0}^{(k)}$, we have

$$A_{n_0}^{(k)} = q\,q_1^{(k-1)}, \quad B_{n_0}^{(k)} = p\,q_{R-1}^{(k-1)}. \tag{2.33}$$

From (2.32) the continuous-time results follow at once. Let $N(t)$ be the process with frequency ν and probabilities p and q for increments $+1$ and -1 respectively. If $q_n(t)$ is the probability that $N(t)$ has not left the interval set $\{1,2,\ldots,R-1\}$ in $(0,t)$ and that $N(t) = n$, we have $q_n(t) = \sum_{k=0}^{\infty} \{(\nu t)^k e^{-\nu t}/k!\}\, q_n^{(k)}$. Hence from (2.32)

$$q_n(t) = 2R^{-1}(p/q)^{\frac{1}{2}(n-n_0)} \times \tag{2.34}$$

$$\times \sum_{\alpha=1}^{R} \sin(n\pi\alpha/R)\sin(n_0\pi\alpha/R)\exp[-\nu t\{1-\sqrt{4pq}\cos(\pi\alpha/R)\}].$$

The passage-time densities $\alpha_{n_0}(\tau)$ and $\beta_{n_0}(\tau)$ for departure at states $n = 0$ and $n = R$ respectively are given by

$$\alpha_{n_0}(\tau) = \nu q\, q_1(\tau); \quad \beta_{n_0}(\tau) = \nu p\, q_{R-1}(\tau). \tag{2.35}$$

IV.3 A generalization of the method of images

The method of images of Section 2 is based on the overt symmetry of equation (2.1), or on the symmetry available through transformation. Even when there is no symmetry, overt or available, a method of images may be exhibited for homogeneous lattice processes when for at least one direction, only unit increments are permitted. This generalized procedure is based on the fact that the Green's functions $g_n^{(k)}$ (or $g_n(t)$) are not linearly independent. The discrete-time functions satisfy the set of equations indexed by $R = 1,2,\ldots$

$$\sum_{n=-\infty}^{\infty} n\, e_n^{(R)}\, g_{-n}^{(k)} = 0, \quad \text{for all } k \geqslant 0. \tag{3.1$_R$}$$

This equation follows directly from $\oint (d/du)\{\epsilon^{R+k}(u)\}\, du = 0$, the contour being about the unit circle, and the fact that $\epsilon(u)$ is a single-valued function there. Then $\oint (d/du)\{\epsilon^R(u)\}\, \epsilon^k(u)\, du = 0$. From $(d/du)\, \epsilon^R(u) = \sum_{-\infty}^{\infty} n e_n^{(R)} u^{n-1}$, and $\epsilon^k(u) = \sum_{-\infty}^{\infty} g_n^{(k)} u^n$ we obtain (3.1$_R$).

When (3.1$_R$) is multiplied by $\{e^{-\nu t}(\nu t)^k/k!\}$ and summed over k, one obtains via (II.4.12) the corresponding continuous-time relations, for $R = 1,2,\ldots$

$$\sum_{n=-\infty}^{\infty} n\, e_n^{(R)}\, g_{-n}(t) = 0 \quad \text{for all } t \geqslant 0. \tag{3.1$'_R$}$$

Another important relation between the Green's function elements is obtained from the equation $\oint \frac{d}{du}\{u^{-n}\, \epsilon^k(u)\}\, du = 0$, or equivalently from

$$\oint \Bigg[-n u^{-n-1} \Big\{ \sum_j g_j^{(k)} u^j \Big\} + \\ + u^{-n} k \Big\{ \sum_j g_j^{(k-1)} u^j \Big\} \Big\{ \sum_m m e_m u^{m-1} \Big\} \Bigg] du = 0$$

from which one obtains, for all $k \geqslant 0$,

$$n g_n^{(k)} = k \sum_m m\, e_m\, g_{n-m}^{(k-1)}. \tag{3.2}$$

If we multiply by $\{e^{-\nu t}(\nu t)^k/k!\}$ and sum over k from 0 to ∞ we find

$$n g_n(t) = \nu t \sum_m m\, e_m\, g_{n-m}(t). \tag{3.2$'$}$$

Consider now the homogeneous processes in discrete time with unit increments only in the positive direction, i.e., for which $e_1 \neq 0$, and $e_n = 0$ for $n > 1$. Consider first the process commencing at $n_0 = 1$, and let $q_n^{(k)}$ be the probability that $N_k = n$ and that the state $n = 0$ has not been reached at any of the first k steps. Then as in Section 2, we have for the set of probabilities $\{q_n^{(k)}\}$, $n = 0,1,2,\ldots$,

$$q_0^{(k)} = 0, \quad (3.3) \qquad\qquad q_n^{(0)} = \delta_{n,n_0} \tag{3.4}$$

and

$$q_n^{(k)} = \sum_1^{\infty} q_{n'}^{(k-1)}\, e_{n-n'}, \qquad n \geqslant 1,\ k \geqslant 1. \tag{3.5}$$

The set $\{q_n^{(k)}\}$ is fully determined from equations (3.4) and (3.5), by recursion on k. The probabilities $q_n^{(k)}$ may now be exhibited in terms of the Green probabilities, with the aid of two observations. The first is that if $\{q_n^{(k)}: n = -1, -2, \ldots\}$ is a set of unspecified auxiliary quantities, (3.5) is satisfied if the augmented set $\{q_n^{(k)}: n = 0, \pm 1, \pm 2, \ldots\}$ satisfies

$$q_n^{(k)} = \sum_{-\infty}^{\infty} q_{n'}^{(k-1)}\, e_{n-n'}. \tag{3.6}$$

This follows from (3.3) and $e_n = 0$ for $n > 1$, provided the series converges. The second observation is that (3.6) is identical in form with the equations $g_n^{(k)} = \sum_{-\infty}^{\infty} g_{n'}^{(k-1)} e_{n-n'}$ for the Green probabilities. Hence (3.6) is satisfied by the set $\{g_{n+j}^{(k)}\}$ for any j and therefore by any linear combination $\Sigma \beta_j\, g_{n+j}^{(k)}$. In particular, a set of coefficients β_j is provided by (3.1_R) with $R = 1$ for which this linear combination vanishes for $n = 0$ for all $k \geqslant 0$ and for which (3.4) is satisfied for positive n. We have in fact as the solution to (3.6) and $q_n^{(0)} = \delta_{n,1}$ for $n > 0$,

$$q_n^{(k)} = g_{n-1}^{(k)} - \sum_1^\infty \frac{j\,e_{-j}}{e_1} g_{n+j}^{(k)} . \tag{3.7}$$

The convergence of the series is assured by (3.2), with whose aid the answer takes the simple form

$$q_n^{(k)} = \frac{1}{e_1} \frac{n}{k+1} g_n^{(k+1)} . \tag{3.8}$$

The passage-time probability is, therefore,

$$s_1^{(k)} = \sum_{n=1}^\infty q_n^{(k-1)} \left\{ \sum_{-\infty}^{-n} e_j \right\} = \frac{1}{e_1 k} \sum_{n=1}^\infty n\, g_n^{(k)} \left\{ \sum_{-\infty}^{-n} e_j \right\} . \tag{3.9}$$

When the negative increments are bounded, the series terminates.

For the associated process $N(t)$ in continuous time (Section II.2) with increment frequency ν, we have from (3.8) for $q_n(t) = \Sigma\{(\nu t)^k \times e^{-\nu t}/k!\} q_n^{(k)}$

$$q_n(t) = \frac{1}{e_1} \frac{n}{\nu t} g_n(t) \tag{3.8$'$}$$

where $g_n(t)$ is the Green probability for the process in continuous time. The passage-time density is therefore

$$s_1(\tau) = \frac{1}{\tau} \sum_{n=1}^\infty \frac{1}{e_1} \left\{ \sum_{-\infty}^{-n} e_j \right\} n\, g_n(\tau) \tag{3.9$'$}$$

The procedure can be generalized to arbitrary initial state n_0, by working with (3.1$_R$) for $R = 1,2,\ldots,n_0$. By eliminating $g_{-1}^{(k)}, g_{-2}^{(k)}, \ldots, g_{-(n_0-1)}^{(k)}$ from these equations, we will obtain a relation between $g_{-n_0}^{(k)}$ and $\{g_j^{(k)},\ j = 1,2,\ldots\}$ of the form

$$g_{-n_0}^{(k)} = -\sum_{j=1}^\infty \alpha_{n_0,j}\, g_j^{(k)} , \tag{3.10}$$

and have, in consequence,

$$q_n^{(k)} = g_{n-n_0}^{(k)} + \sum_1^\infty \alpha_{n_0,j}\, g_{n+j}^{(k)} . \tag{3.11}$$

For $k=0$, $q_n^{(0)} = \delta_{n,n_0} + \sum_1^\infty \alpha_{n_0,j}\,\delta_{n,-j}$ satisfying (3.4). The set of numbers $\alpha_{n_0,j}$ has the character of an image configuration, mirroring an initial concentration at $n = n_0$ so as to give rise to $q_0^{(k)} = 0$ for all k. The passage-time probability for n_0 is

$$s_{n_0}^{(k)} = \sum_{n=1}^{\infty} \left\{ \sum_{j=n}^{\infty} e_{-j} \right\} q_n^{(k-1)}. \tag{3.12}$$

When negative increments are bounded, the series (3.12) terminates. If, for example, $e_n = e_1\,\delta_{n,1} + e_{-K}\,\delta_{n,-K}$, then $s_{n_0}^{(k)} = e_{-K} \sum_{n=1}^{K} q_n^{(k-1)}$.

The extension to continuous time is immediate. For processes $N(t)$ with increment frequency ν, $e_1 \neq 0$ and $e_n = 0$ for $n > 1$, the probability $q_n(t) = P\{N(t) = n,\ N(t') > 0 \text{ for } 0 \leqslant t' \leqslant t \,|\, N(0) = n_0\}$ is $\Sigma\, q_n^{(k)} \times \{e^{-\nu t}(\nu t)^k/k!\}$, i.e.,

$$q_n(t) = g_{n-n_0}(t) + \sum_1^{\infty} \alpha_{n_0,j}\, g_{n+j}(t) \tag{3.13}$$

and

$$s_{n_0}(\tau) = \nu \sum_{n=1}^{\infty} \left\{ \sum_{j=n}^{\infty} e_{-j} \right\} q_n(\tau). \tag{3.14}$$

Again the series terminates when negative increments are bounded. Because (3.13) involves the Green's functions the asymptotic approximation methods of Chapter III, based on the Central Limit Theorem and conjugate transformation, are available.

IV.4 The adjoint system: unit negative increments

When the process permits unit increments in the negative direction, the procedure of Section 3 must be modified. Again we consider the process in discrete time first.

Let $s_{n_0}^{(k)}$ be the probability that the process N_k commencing at n_0 attains the value $n = 0$ for the first time at the kth step. These probabilities are related by a simple recursion law with structure similar to that governing the process of the previous section. Specifically we have

$$s_{n_0}^{(k)} = \sum_m e_m\, s_{n_0+m}^{(k-1)}, \qquad k \geqslant 2, \tag{4.1}$$

provided that $s_0^{(k)}$ is understood to be zero for all $k \geqslant 1$. For if one commences at $n = n_0$, a first step takes one to $n_0 + m$, and a return to $n = 0$ in $(k-1)$ steps is then required to contribute to $s_{n_0}^{(k)}$. If the state $n = 0$ is reached at the first step, the contribution to $s_{n_0}^{(k)}$ for $k \geqslant 2$ is zero. For $k = 1$, we have

$$s_{n_0}^{(1)} = e_{-1}\,\delta_{n_0,1}. \tag{4.2}$$

Equation (4.1) may be rewritten as

$$s_{n_0}^{(k)} = \sum_{j=1}^{\infty} \tilde{e}_{n_0-j}\, s_j^{(k-1)}, \qquad k \geqslant 2, \tag{4.3}$$

where $\tilde{e}_n = e_{-n}$, i.e., where $\tilde{e}_n$ is the increment probability for the *adjoint process* with positive and negative directions interchanged. We may now compare (4.2) and (4.3) with (3.4) and (3.5) and infer that $s_n^{(k)} = e_{-1}\,\tilde{q}_{1n}^{(k-1)}$, where $\tilde{q}_{1n}^{(k-1)}$ is the $q_n^{(k-1)}$ of Section 3 for the adjoint process commencing at $n_0 = 1$. Hence we have from (3.8)

$$s_{n_0}^{(k)} = \frac{n_0}{k}\,\tilde{g}_{n_0}^{(k)} = \frac{n_0}{k}\,g_{-n_0}^{(k)}. \tag{4.4}$$

The result (4.4) has been given by Kemperman (1961, p. 47, (7.15)). The result for the special case $e_{-1} + e_K = 1$, K a positive integer, was given earlier by Korolyuk in 1955 and Blackman in 1956, as stated by Kemperman.†

For the processes $N(t)$ in continuous time of frequency ν, for which only unit negative increments are permitted, we have at once, from the reasoning for (2.13), that

$$s_{n_0}(\tau) = \frac{n_0}{\tau}\,g_{-n_0}(\tau). \tag{4.5}$$

where $s_{n_0}(\tau)$ is the density of passage times τ from the state $n = n_0$ to the state $n = 0$.

IV.4.1 Application to bulk queues

Let us illustrate the results of Sections 3 and 4, by applying them to two distinct single-server queuing systems. The first system has a bulk service feature and is defined in the following way.

† A comparison of the methods and results of this chapter with those of Kemperman is given in Report S323 of the Mathematics Centre in Amsterdam by J. Runnenberg.

(1) Passengers arrive one at a time to a bus queue at Poisson epochs of rate λ. Buses arrive at other independent Poisson epochs of rate η. The buses accept K passengers or any smaller number in queue. The queue persistence-time τ may be defined as the length of a queuing period initiated by the arrival of a customer to an empty queue and terminated by the depletion of the queue. The queue persistence-time density is clearly identical with the density $s_1(\tau)$ for the homogeneous process $N(t)$ commencing at $N(0) = 1$ of passage times to the non-positive states. The homogeneous process has increment frequency $\nu = \lambda + \eta$, and increment probabilities $e_n = (\lambda/\nu)\delta_{n,1} + (\eta/\nu)\delta_{n,-K}$ (Section II.2.1). We may then employ (3.9′) to find

$$s_1(\tau) = \frac{1}{\tau}\left(\frac{\eta}{\lambda}\right)\sum_{1}^{K} n\, g_n(\tau), \tag{4.6}$$

where the Green probability $g_n(t) = (2\pi i)^{-1} \oint u^{-(n+1)} \exp\{-\nu t(1 - e_1 u - e_{-K} u^{-K})\} du$ is a generalized Bessel function as described in Section II.9. The result (4.6) is in agreement with that obtained by a different method (Keilson, 1964a).

(2) In a second type of single-server queuing system, customers arrive in groups of random group-size and are served individually. Let the groups arrive at Poisson epochs of rate λ and let the group-size be m with probability c_m ($\Sigma c_m = 1$). Let the time spent by a customer in service be an independent random variable with exponential density $f(x) = \eta e^{-\eta x} U(x)$. If at $t = 0$, there are n_0 customers in the system (in line and in service), the server will remain occupied for a period of duration τ which terminates when the number of customers in the system reaches the value zero. The server occupation-time density $s_{n_0}(\tau)$ is identical with the passage-time density for the homogeneous process $N(t)$ with frequency $\nu = \lambda + \eta$ and increment probability $e_n = (\lambda/\nu)c_n + (\eta/\nu)\delta_{n,-1}$, commencing at $N(0) = n_0$. From (4.5) we then have

$$s_{n_0}(\tau) = \frac{n_0}{\tau}\, g_{-n_0}(\tau) \tag{4.7}$$

where $g_n(\tau) = (2\pi i)^{-1} \oint u^{-(n+1)} \exp\{-\nu\tau(1 - \Sigma e_n u^n)\} du$ is a generalized Bessel function as described in Section II.9.

IV.5 Skip-free processes on the continuum

Consider the process

$$X(t) = X_0 + vt + X_1(t) + X_2(t), \tag{5.1}$$

where $X_1(t)$ is the Wiener-Lévy process with zero mean and variance $\sigma_1^2(t) = \alpha_1 t$ (Section II.3d) and $X_2(t)$ is a homogeneous jump process (Section II.3b) with positive increments of distribution $A(x)$ and frequency ν. The samples of the Wiener process $X_1(t)$ are continuous in time (Section II.7), and the samples of the process $X_2(t)$ are monotonic increasing step functions. Consequently, it is clear that the net process $X(t)$ has the property of being skip-free in the negative direction in the sense of Section 1. Some examples of this type of process have been given in Section II.3. Other examples are described in Section V.1.

It may be anticipated from the basic result (4.5) of the previous section that a similar result will be available for the process (5.1), which may be regarded as a limiting form of the lattice processes. Rather than obtain such a result by a limiting procedure, a direct analytical derivation will be given.

An immediate consequence of the skip-free property of the motion is the following classical renewal argument. By virtue of the diffusive component $X_1(t)$, the distribution $G(x,t)$ of $X(t)$ will be absolutely continuous for $t > 0$. If $X(0) = y$, the process $X(t)$ will have the probability density $g(x-y, t) = (d/dx)P\{X(t) \leqslant x|X(0) = y\}$ where $g(x,t)$ is the Green density. Let the distribution of the passage time τ from y to zero be denoted by $S_y(\tau) = P\{X(t') = 0$ for some t' in $(0,\tau)|X(0) = y\}$. A process sample commencing at $y > 0$ will be found in a differential interval about $-x < 0$ at time t only if the process assumed the value $x = 0$ one or more times in $(0,t)$, and if at a time $(t-\tau)$, subsequent to the first such time τ, it assumed a value in that differential interval about $-x$. Hence for $x > 0$, and $y > 0$, we have

$$g(-x-y,t) = \int_0^t g(-x, t-\tau)\,dS_y(\tau). \tag{5.2}$$

If we denote the Laplace-Stieltjes transform of $G(x,t)$ by $\gamma^*(x,s)$ and the Laplace-Stieltjes transform of $S_y(\sigma)$ by $\sigma_y^*(s)$, we obtain† from (5.2)

$$\sigma_y^*(s) = \frac{\gamma^*(-x-y, s)}{\gamma^*(-x,s)} \quad \text{for } x > 0. \tag{5.3}$$

The function $\gamma^*(x,s)$ may be obtained as follows. From II.7.3 the charac-

† See footnote on next page.

teristic function of $X(t)$ for $y = 0$ is $\mathfrak{F}_x\{g(x,t)\} = \exp[-Dz^2t + izvt - \nu t\{1-\alpha(z)\}]$ and the Laplace transform $\mathfrak{L}_t\mathfrak{F}_x\{g(x,t)\} = [s + Dz^2 - izv + \nu\{1-\alpha(z)\}]^{-1}$. We have therefore, for $\gamma^*(x,s) = \mathfrak{L}_t\{g(x,t)\}$, the inverse Fourier transform

$$\gamma^*(x,s) = \frac{1}{2\pi}\int_{-\infty}^{\infty} e^{-izx}[s + Dz^2 - izv + \nu\{1-\alpha(z)\}]^{-1}dz. \qquad (5.4)$$

When $\mathrm{Re}(s) > 0$, the integral is absolutely convergent, $\gamma^*(x,s)$ is analytic in s and vanishes as $|s| \to \infty$. The distribution of increments $A(x)$ has positive support only, so that its c.f.$\alpha(z)$ is analytic in the upper half-plane, and goes to zero there as $|z| \to \infty$. When $x < 0$, the contour may be closed in the upper half-plane and the integral (5.4) exhibited in terms of the residues there. The integrand may be seen to have precisely one simple zero when $\mathrm{Re}(s) > 0$, $\mathrm{Im}(z) \geqslant 0$. For consider the function

$$\psi(z,s) = 1 - \{\nu/(\nu + s - izv + Dz^2)\}\,\alpha(z). \qquad (5.5)$$

When z is real and $\mathrm{Re}(s) > 0$, the expression in curly brackets is of magnitude less than unity and $|\alpha(z)| \leqslant 1$ so that $\mathrm{Re}\{\psi(z,s)\} > 0$. At infinity in the upper half-plane, $\psi(z,s) = 1$. From the Principle of the Argument we infer that $\psi(z,s)$ has as many zeros as singularities in the upper half-plane. Since $\alpha(z)$ is analytic there and $(\nu + s - izv + Dz^2)$ has precisely one zero there, as may readily be seen, $\psi(z,s)$ and the denominator of the integrand in (5.4) have one simple zero there. If we denote this zero by $z = \zeta(s)$ and the residue for $x = 0$ at this zero by $\rho(s)$, we have from (5.4),

---

† (See previous page)
Equation (5.3) has a direct analogue for the skip-free lattice processes with unit increments in the negative direction. By the same argument we have in the notation of Section 4 for the processes in discrete time, $g^{(k)}_{-n_0-n} = \sum_{k'=1}^{k} s^{(k')}_{n_0}\, g^{(k-k')}_{-n}$, and for the corresponding generating functions with respect to k, $\sigma_{n_0}(w) = \gamma_{-n_0-n}(w)/\gamma_{-n}(w)$. For the associated process in continuous time (Section II.2), $g_{-n_0-n}(t) = s_{n_0}(t) * g_{-n}(t)$, so that

$$\sigma^*_{n_0}(s) = \frac{\gamma^*_{-n_0-n}(s)}{\gamma^*_{-n}(s)}. \qquad (5.3')$$

$$\gamma^*(-y, s) = i\rho(s) \exp\{iy\,\zeta(s)\}. \tag{5.6}$$

Equation (5.3) then becomes

$$\sigma_y^*(s) = \exp\{iy\,\zeta(s)\}. \tag{5.7}$$

It follows that for $a,b > 0$, $\sigma_{a+b}^*(s) = \sigma_a^*(s)\,\sigma_b^*(s)$ as is required from the skip-free character of the process. The residue at $z = \zeta(s)$ for $x = 0$ is given from L'Hôpital's rule by

$$\begin{aligned}\rho(s) &= \lim_{z\to\zeta(s)} (z-\zeta(s))/[s + Dz^2 - izv + \nu\{1-\alpha(z)\}] \\ &= [-iv + 2D\,\zeta(s) - \nu\,\alpha'(\zeta(s))]^{-1}. \end{aligned} \tag{5.8}$$

We also have, since $\zeta(s)$ is the zero of the denominator, $s - i\zeta(s)v + D\zeta^2(s) + \nu\{1-\alpha(\zeta(s))\} = 0$. If we differentiate the latter equation with respect to s, we find that $1 + [2D\,\zeta(s) - iv - \nu\,\alpha'(\zeta(s))]\,\zeta'(s) = 0$, whence

$$\rho(s) = -\zeta'(s). \tag{5.9}$$

Introducing (5.9) into (5.6) we have $\gamma^*(-y,s) = -y^{-1}\frac{d}{ds}[\exp\{iy\,\zeta(s)\}] = -y^{-1}\frac{d}{ds}\sigma_y^*(s)$, so that $\sigma_y^*(s) = y\int_s^\infty \gamma^*(-y,s')\,ds'$. Reference to a table of Laplace Transforms gives the required result:†

$$s_y(\tau) = \frac{y}{\tau}g(-y,\tau). \tag{5.10}$$

When $D = 0$, $G(x,t)$ is not absolutely continuous, nor is the passage-time distribution $S_y(\tau)$ continuous in τ. Equation (5.10), however, continues to be true in a generalized sense. If, for example, $A(x)$ is absolutely continuous, then we may write, from (II.3.12),

$$\frac{y}{t}g(-y,t) = \frac{y}{t}\,\delta(-y-vt)\,e^{-\nu t} + \frac{y}{t}\sum_1^\infty\left\{\frac{(\nu t)^k}{k!}\,e^{-\nu t}\right\} a^{(k)}(-y-vt).$$

Hence from (5.9),

$$s_y(\tau) = \delta(\tau - |y/v|)e^{-\nu|y/v|} + \frac{y}{\tau}\sum_1^\infty \frac{(\nu\tau)^k}{k!}\,e^{-\nu\tau}\,a^{(k)}(|v|\tau - y). \tag{5.11}$$

---

† This result was first obtained by D.G. Kendall (1957) in a slightly different form. The result is discussed in Keilson (1963a).

IV.6 Connection with other results

For the pure diffusion process to which (5.1) reduces when $X_2(t) = 0$, the Green density is given by (II.8.1) to be $g(x,t) = (4\pi Dt)^{-\frac{1}{2}} \times \exp[-(x-vt)^2/4Dt]$. Consequently, for $s_y(\tau)$, the density of passage-times τ from y to 0, we have from (4.5)

$$s_y(\tau) \;=\; \frac{y}{\tau}(4\pi D\tau)^{-\frac{1}{2}} \exp\left[-(y+v\tau)^2/4D\tau\right], \tag{6.1}$$

and this is a classical result in diffusion theory.

When the diffusive component $X_1(t)$ of (5.1) is zero, the process reduces to the discontinuous drift process $X_v(t)$ given by (II.3.11). The process for $v = -1$ arises in the study of a single-server queuing system said to be of the $M/G/1$ type (Kendall, 1953). The system is governed by the following rules:

Customers arrive individually at Poisson epochs of frequency ν, and are served one at a time. The service times required for successive customers are independent, and have a common distribution $A(x)$. Service is provided without interruption.

When the mean time between arrivals ν^{-1} is greater than the mean service time $\int x\,dA(x)$, the events at which the server becomes idle are recurrent, and periods in which the server is busy alternate with periods of idleness. The durations of the busy periods constitute a set of independent identically distributed random variables. The distribution $S(\tau)$ of the server busy period was shown by D.G. Kendall (1951) to be characterized by a transcendental equation governing its Laplace-Stieltjes transform $\sigma^*(s) = \mathfrak{L}-\mathfrak{S}\{S(\tau)\}$, i.e.,

$$\sigma^*(s) \;=\; \alpha^*(s+\nu-\nu\sigma^*(s)) \tag{6.2}$$

where $\alpha^*(s)$ is the Laplace-Stieltjes transform of the service-time distribution $A(x)$. We wish to relate this equation to the basic result (5.10) of the previous section.

The total service load arriving at the system in the interval $(0,t)$ is described statistically by a jump process $X_2(t)$ governed by the increment distribution $A(x)$ and increment frequency ν. If at $t = 0$ the server is idle and a customer arrives with the service-time requirement y, the time τ_y for which the server will remain busy will have a distribution $S_y(\tau) = P\{\tau_y \leqslant \tau\} = P\{y + X_2(t') - t' = 0 \text{ for some } t' \text{ in } [0,\tau]\}$. Hence $S_y(\tau)$ is the passage-time distribution of the process (5.1) with $X_0 =$

y, $X_1(t) = 0$ and $v = -1$. Since the distribution of initial server backlog y is $A(y)$, the busy-period distribution as defined above will be given by

$$S(\tau) = \int S_y(\tau)\, dA(y). \tag{6.3}$$

We now wish to show that (6.3) with $S_y(\tau)$ obtained from (5.10) satisfies (6.2).

Consider the expression (6.3) for processes (5.1) with $D \neq 0$. Since $\mathfrak{L}$ - $\mathfrak{S}\{S_y(\tau)\} = \exp\{iy\zeta(s)\}$ by (5.7), $\sigma^*(s) = \mathfrak{L}$ - $\mathfrak{S}\{S(\tau)\} = \int e^{iy\,\zeta(s)}dA(y)$, i.e.,

$$\sigma^*(s) = \alpha(\zeta(s)). \tag{6.4}$$

The zero $\zeta(s)$ in the upper half-plane of $\psi(z,s)$ in (5.5) approaches a limiting zero in the upper half-plane as $D \to 0$. This follows again from Rouché's theorem and the observation that $\nu + s + iz$ has one zero there for $\mathrm{Re}(s) > 0$. Hence $\zeta(s) = i\{s + \nu - \nu\alpha(s))\}$, and since $\alpha(is) = \alpha^*(s)$, (6.2) follows from (6.4).

The characterization (6.2) is of somewhat limited practical value. Moments may be obtained from it by differentiation. For the study of the asymptotic behaviour of the passage-time densities, employment of (5.10) and the central limit techniques of Chapter III is indicated.

IV.7 The asymptotic behaviour of the passage-time densities; the tail; the diffusion approximation

The passage-time densities discussed in Sections 3 to 5 are simple linear combinations of the Green's functions. Consequently, the asymptotic behaviour of the passage-time densities may be inferred from the asymptotic behaviour of the Green's functions exhibited in Chapter III. We will discuss, as typical, the skip-free processes on the continuum of Section 5, with $D > 0$, where one has the simple relation for the passage-time density from $y > 0$ to negative states given by

$$s_y(\tau) = \frac{y}{\tau} g(-y,\tau). \tag{7.1}$$

The saddlepoint approximation (III.7.11) may be employed when detailed numerical information is desired. When the structure of the asymptotic behaviour is desired, the coarser Gaussian approximations of (III.7.12) are of greater interest. Two distinct domains of asymptotic behaviour are important. The first domain, which may be called the tail of the passage-time density, is described by an asymptotic representation of $s_y(\tau)$ as $\tau \to \infty$. The conjugate density $g_V(x,t)$ appropriate to the description of this tail has $\mu_V(t) = 0$. It is easy to see that such a density will always be available. From (III.7.12) for $V = V_0$, and $\mu_{V_0}(t) = 0$ (cf. Note 6), we have the asymptotic representation $\tau \to \infty$,

$$s_y(\tau) \sim \frac{y}{\tau} e^{-V_0 y} \gamma(iV_0,\tau)\{2\pi\sigma_{V_0}^2(\tau)\}^{-\frac{1}{2}} \exp\{-y^2/2\sigma_{V_0}^2(\tau)\}, \qquad (7.2)$$

where $\gamma(iV,\tau) = \exp[-\nu\tau\{1-\alpha(iV)\} + DV^2\tau + V\tau]$ and $\sigma_V^2(\tau) = 2D\tau + \nu\tau\int x^2 e^{-Vx} dA(x)$. The approximation in (7.2) will be valid when (cf. Note 6)

$$\nu_{V_0}\tau \gg 1; \text{ and } \{(\nu_{V_0}\tau)^{2/3}\sigma_{V_0}(\nu_{V_0}^{-1})\}^{-1} y \ll 1. \qquad (7.3)$$

The second domain of asymptotic behaviour is that associated with the Central Limit Theorem, and gives rise to the *diffusion approximation*. This is associated with the value $V = 0$, and is given from (III.7.12) by

$$s_y(\tau) \approx \frac{y}{\tau}\{2\pi\sigma_0^2(\tau)\}^{-\frac{1}{2}} \exp\{-\{y+\mu_0(\tau)\}^2/2\sigma_0^2(\tau)\} = d_y(\tau) \qquad (7.4)$$

when $\alpha(z)$ is analytic at $z = 0$. The approximation (7.4) will be valid when

$$\nu\tau \gg 1; \qquad \frac{|y + \mu_0(\tau)|}{(\nu\tau)^{2/3}\sigma_0(\nu^{-1})} \ll 1. \qquad (7.5)$$

When y is close to the absorbing boundary at $x = 0$, the set of passage times τ for which this approximation to the passage-time density is useful may have low probability. As we will see, however, when y moves away from the boundary, the diffusion approximation becomes dominant, as described by the following theorem.

Theorem IV.7.1. Let the Green density $g(y,t)$ of (7.1) have mean $\mu(t) = -\bar{\mu}t < 0$. Let $I(y,\gamma)$ be the set of values τ for which the relative error in the diffusion approximation (7.4) is smaller than γ, i.e., for which $|s_y(\tau) - d_y(\tau)| / d_y(\tau) \leqslant \gamma$. Let $p_y(\gamma) = \int_{I(y,\gamma)} s_y(\tau)\,d\tau$. Then for any $\gamma > 0$, one has

$$\lim_{y\to\infty} p_y(\gamma) = 1. \qquad (7.6)$$

To prove the theorem we will first show that

$$\lim_{y\to\infty} \underbrace{\int_{\alpha_1 y/\bar{\mu}}^{\alpha_2 y/\bar{\mu}} t^{-1} y\, g(-y,t)\,dt}_{Q(y)} \;=\; 1, \qquad \text{if } 0 < \alpha_1 < 1 < \alpha_2 . \tag{7.7}$$

For let $g(-y,t) = g_0(\bar{\mu}t - y, t)$, and let $\bar{\mu}t - y = (y/\bar{\mu})^{\frac{1}{2}} s$. Then

$$Q(y) \;=\; \int_{(\alpha_1-1)(\bar{\mu}y)^{\frac{1}{2}}}^{(\alpha_2-1)(\bar{\mu}y)^{\frac{1}{2}}} \left(\frac{y}{\bar{\mu}}\right)^{\frac{1}{2}} g_0\left((y/\bar{\mu})^{\frac{1}{2}} s,\ \frac{y + (y/\bar{\mu})^{\frac{1}{2}} s}{\bar{\mu}}\right) \frac{y}{y + (y/\bar{\mu})^{\frac{1}{2}} s}\, ds. \tag{7.8}$$

The density $g_0(x,1)$ has zero mean. If we denote its variance by σ^2, the Central Limit Theorem implies that $\zeta^{\frac{1}{2}} g_0(\zeta^{\frac{1}{2}}\psi_\zeta, \zeta) \to (2\pi\sigma^2)^{-\frac{1}{2}} \times \exp\{-\psi^2/2\sigma^2\}$, when $\zeta \to \infty$, provided that $\psi_\zeta \to \psi$.† We see therefore that, as $y \to \infty$, the integrand of (7.8) approaches $(2\pi\sigma^2)^{-\frac{1}{2}} \times \exp\{-s^2/2\sigma^2\}$ for fixed s and the limits of integration simultaneously approach $\pm\infty$. Hence $1 \leqslant \liminf Q(y)$ from Fatou's Lemma. The integrand of (7.7) is a probability density, whence $Q(y) < 1$, $\limsup Q(y) \leqslant 1$, and $\lim_{y\to\infty} Q(y) = 1$, verifying (7.7).

We note, in the same way, that (7.7) is still true when $\alpha_1 = \alpha_1(y) = 1 - Ay^{-\frac{1}{2}+\epsilon}$ and $\alpha_2 = \alpha_2(y) = 1 + Ay^{\frac{1}{2}+\epsilon}$ for any $A > 0$ and any small positive ϵ. Let J_y be the closed interval $[\alpha_1(y), \alpha_2(y)]$, and let the relative error be denoted by $E_y(\tau) = |s_y(\tau) - d_y(\tau)| / d_y(\tau)$. Consider the function

$$\Gamma(y) \;=\; \max\{E_y(\tau) \,|\, \tau \in J_y\}. \tag{7.9}$$

For any sequence $\tau = \tau_y$ in J_y, $\bar{\mu}\tau_y/y \to 1$ as $y \to \infty$. Moreover from Theorem III.7.3 of Note 6, $g(-y,t)$ is asymptotically normal in J_y and $s_y(\tau_y) \sim d_y(\tau_y)$. We then infer from (7.9) that $\Gamma(y) \to 0$ as $y \to \infty$.

By definition, $I(y,\gamma)$ is the set of values τ for which $E_y(\tau) \leqslant \gamma$. Since $\Gamma(y) \to 0$, $\max\{E_y(\tau) \,|\, \tau \in J_y\} < \gamma$ for all y greater than some y_γ. Hence the interval $J_y \subset I(y,\gamma)$, for $y > y_\gamma$. Consequently

$$1 \geqslant \int_{I(y,\gamma)} s_y(\tau)\,d\tau \geqslant \int_{J_y} s_y(\tau)\,d\tau \qquad \text{for all } y > y_\gamma . \tag{7.10}$$

Since we have seen that the right side of (7.10) $\to 1$ when $y \to \infty$, $\int_{I(y,\gamma)} s_y(\tau)\,d\tau$ also goes to 1. Q.E.D.

---

† For $\zeta^{\frac{1}{2}} g_0(\zeta^{\frac{1}{2}} x, \zeta)$ goes to the normal density uniformly in x on $(-\infty,\infty)$.

CHAPTER V

BOUNDED PROCESSES AND ERGODIC GREEN'S FUNCTIONS

A variety of bounded processes arising in applied probability theory may be regarded as homogeneous processes modified by the presence of boundaries. A systematic pursuit of this point of view gives rise to the introduction of certain compensation functions which represent the effect of the boundaries as source distributions for the homogeneous process. For ergodic processes these source distributions become invariant with time and the time-dependent Green's functions become replaced by certain ergodic Green's functions. The resulting formalism provides an effective method for studying the bounded processes in the real domain as an alternative to the more formal Wiener-Hopf and Hilbert problem methods generally employed. The character of the ergodic Green's functions for the various types of homogeneous processes encountered, and the convergence of their defining integrals and series are described.

V.1 Bounded homogeneous processes

Many processes of theoretical and practical interest may be loosely described as bounded processes which are spatially homogeneous away from the boundaries. A simple but important example is the process

$$X_k = \max[0, X_{k-1} + \xi_k], \qquad k = 1,2,\ldots \tag{1.1}$$

where X_0 is a prescribed random variable, and the increments ξ_k are a sequence of independent and identically distributed random variables. The process (1.1) is bounded from below, i.e., every process sample is non-negative for all k. If the increments have a distribution $A(x)$, the single-step transition distribution $A(x',x)$ (Section II.2) corresponding to (1.1) is given for $0 \leqslant x'$ by

$$\begin{aligned} A(x',x) &= 0 & x &< 0 \\ A(x',x) &= A(x-x') & x &\geqslant 0\,. \end{aligned} \tag{1.2}$$

The process of (1.1) may be regarded as being a spatially homogeneous process $X_k^* = X_{k-1} + \xi_k$, modified by the presence of a *retaining*† *boundary* at the state $x = 0$. The word "retaining" is used in the sense of holding back. If a process sample has the value x_{k-1} after the $(k-1)$st step and the next increment is ξ_k, the next value for the homogeneous process would be $x_{k-1} + \xi_k$. If this value is negative, the boundary acts to replace it by the value zero, the smallest permissible value for the process. The incidence of the process (1.1) in practice is illustrated by the following example.

(1) *A waiting-time process in queuing theory.* A stream of customers arrive at a single-server queue at a sequence of arrival epochs τ_0, τ_1, etc., the kth customer arriving at epoch τ_k. The interarrival times $T_k = \tau_{k+1} - \tau_k$ are independent and have a common distribution $T(x)$. The service times S_k required by the kth customer form a separate sequence of independent random variables with the common distribution $S(x)$. If X_k is the time the kth customer must wait in queue for service, then $X_{k+1} = \max[0, X_k + \xi_{k+1}]$ where $\xi_{k+1} = S_k - T_k$ (Lindley, 1952). In this context, the process is called the Lindley waiting-time process.

Similarly one may deal with a spatially homogeneous process modified by two retaining boundaries at $x = 0$ and $X = L$. The law $X_k^* = X_{k-1}^* + \xi_k$ governing the process X_k^* in the absence of boundaries is replaced, for the modified process X_k, by

$$
\begin{aligned}
X_k &= 0 && \text{if} \quad X_{k-1} + \xi_k \leqslant 0, \\
X_k &= X_{k-1} + \xi_k && \text{if} \quad 0 < X_{k-1} + \xi_k < L, \\
X_k &= L && \text{if} \quad X_{k-1} + \xi_k \geqslant L.
\end{aligned}
\tag{1.3}
$$

The process consists of the addition of the independent random variables ξ_k with replacement of X_k by the values 0 and L when the boundaries of the interval $(0,L)$ are exceeded. We then have for the single-step transition distribution, for $0 \leqslant x' \leqslant L$,

$$
\begin{aligned}
A(x', x) &= A(x - x') && 0 \leqslant x < L \\
A(x', x) &= 1 && x \geqslant L.
\end{aligned}
\tag{1.4}
$$

† For lattice processes with Bernoulli increments, and diffusion processes, the term *reflecting boundary* is customary. The retaining boundary is also called "impenetrable".

When $L = \infty$, the process (1.3) reduces to (1.1).

A generalization of the process (1.3) of frequent interest is the following *replacement process*. When $X_{k-1} + \xi_k < 0$, the process takes on a random value $X_k = \zeta_{1k}$. The new value ζ_{1k} is an independent random variable with a prescribed distribution $B_1(x)$. Similarly, when $X_{k-1} + \xi_k > L$, $X_k = \zeta_{2k}$ where the ζ_{2k} are independent random variables with a prescribed distribution $B_2(x)$. The distributions $B_1(x)$ and $B_2(x)$ have their support on the interval $[0,L]$. Formally,

$$
\begin{aligned}
X_k &= \zeta_{1k} && \text{if} \quad X_{k-1} + \xi_k \leqslant 0, \\
X_k &= X_{k-1} + \xi_k && \text{if} \quad 0 < X_{k-1} + \xi_k < L, \\
X_k &= \zeta_{2k} && \text{if} \quad X_{k-1} + \xi_k \geqslant L;
\end{aligned}
\tag{1.5}
$$

and for $0 \leqslant x' \leqslant L$,

$$A(x', x) = 0 \qquad x < 0, \tag{1.6a}$$

$$A(x', x) = A(x-x') - A(-x') + A(-x')B_1(x) + \{1 - A(L-x')\}B_2(x), \qquad 0 \leqslant x < L, \tag{1.6b}$$

$$A(x', x) = 1 \qquad x \geqslant L. \tag{1.6c}$$

The process (1.3) is a subcase of (1.5) with $B_1(x) = U(x)$ and $B_2(x) = U(x-L)$. The distributions $B_1(x)$ and $B_2(x)$ will be called *replacement distributions*. The processes on the half-line $[0,\infty)$ with a single retaining boundary at $x = 0$ may be regarded as subcases with $L = \infty$.

The jump processes $X(t)$ in continuous time associated (Section II.2) with the processes (1.1), (1.3) and (1.5) also arise in applications. Two examples follow.

(2) *A Poisson queue with finite capacity, group arrivals, and group service*

Customers arrive at a bus queue in groups at Poisson epochs of frequency ν_1 with group size distribution $D(x)$. Buses arrive at Poisson epochs of frequency ν_2 with random capacity of distribution $C(x)$. The queue has a capacity L, and excess customers leave. The number of customers $X(t)$ in queue is then a process of the type described in Section II.2.1, subject to the two competing classes of events. In the notation of that section, $A_1(x', x) = D(x-x')$ for $0 \leqslant x < L$; $A_1(x', x) = 1$, for $x \geqslant L$; $A_2(x',x) = 1 - C(x'-x)$, $0 \leqslant x < L$; $A_2(x',x) = 0$ for $x < 0$. The process is then equivalent to the jump process in continuous time of frequency $\nu = \nu_1 + \nu_2$ and single-step transition distribution $A(x',x) = (\nu_1/\nu).A_1(x',x) + (\nu_2/\nu).\, A_2(x',x)$ by (II.2.14). $A(x',x)$ has the structure of (1.4).

(3) *An inventory model*

A warehouse is subject to shipments of frequency ν_1 with random size of distribution $D(x)$. Orders arrive at frequency ν_2 with order size distribution $C(x)$. The warehouse has a capacity L, and if a shipment is received exceeding this capacity, the excess is diverted elsewhere. When an order is received in excess of the inventory level, the amount on hand is shipped and the balance of the order is diverted. The structure of the process $X(t)$ describing the inventory level is then identical with that for the queue length above. If instead, the inventory level is restored to the capacity L when depleted, the structure is that of (1.6) with $B_1(x) = U(x-L)$ and $B_2(x) = U(x-L)$.

The homogeneous drift processes and the processes with diffusion components of Section II.3 may also be modified by retaining boundaries. As we have seen in Section II.7, such processes may be regarded as limiting forms of the jump processes. The bounded processes are correspondingly limiting forms of the bounded jump processes, and fall within the framework of their discussion. Two examples follow.

(4) *A dam with stochastic inputs and constant efflux*

Consider a dam with positive increments ξ_i due to rainfall of distribution $A_0(x)$ occurring at Poisson epochs of frequency ν_0. Between these epochs, the water level $X(t)$ decreases at the constant rate $dX/dt = v < 0$ as long as $X(t)$ is positive. When depletion occurs, $X(t)$ remains at zero until the next increment epoch. If the capacity L of the dam is finite, any overflow is lost.

The method of Section II.7B may be called upon. One may work with the jump process in continuous time of frequency

$$\nu_j = \nu_0 + n_j|v| \tag{1.7a}$$

and increment distribution

$$A_j(x) = \nu_j^{-1}\{\nu_0 A_0(x) + \eta_j|v|\, e^{\eta_j x}\, U(-x) + \eta_j|v|\, U(x)\} \tag{1.7b}$$

modified by a retaining boundary at $x = 0$. The modified drift process for the dam may then be treated as the limit of the sequence of modified jump processes $\{\nu_j, A_j(x)\}$ with $\eta_j \to \infty$.

(5) *The service backlog process of Takács (1955)*

A single server provides service to customers (individuals or groups) arriving at Poisson epochs of frequency ν_0. The service time requirement due to an arrival is an independent random variable of distribution $A_0(x)$. The service backlog $X(t)$ is defined as the amount of time the server must expend to complete all service requirements at hand at time t. Between arrivals epochs, $X(t)$ decreases uniformly at the rate $dX(t)/dt = -1$. If the work at hand is depleted, the server waits for the next arrival. The process $X(t)$ is a Markov process with structure identical to that of the process above for a dam of infinite capacity and may be treated in the same manner.

(5a) *The service backlog process of (5) modified by a secondary function of the server*

When the backlog $X(t)$ reaches the value zero, the server accepts a secondary task of duration ζ with distribution $B(x)$, and $X(t)$ is set at the new value ζ. This may be regarded as a replacement process for the homogeneous drift process.

Passage processes terminating when the boundary at $x = 0$ or $x = L$ is traversed for the first time may be of interest. The boundary will then be called an *absorbing boundary*.

For processes with one absorbing boundary, it is generally convenient to augment the state space with a separate completion state. For processes with two absorbing boundaries two completion states are required when the identity of the boundary exceeded is of interest. Such passage processes may be discussed within the framework of the replacement process by setting $B_1(x)$ and/or $B_2(x) = 0$ in (1.6) and modifying (1.6c) suitably. A treatment of passage processes along these lines may be found in Keilson, 1964b. Discussion of passage processes will be resumed in Chapter VII.

V.2 Compensation functions for discrete time processes

A treatment of the homogeneous bounded processes of Section 1 may be centred about the Green's functions of Section II.3 by introducing compensation functions to describe the influence of the inhomogeneities associated with the boundaries. The nature of these compensation functions and the manner in which they enter are the subject of this and the following section.

For the processes in discrete time, the distribution after k steps is obtained recursively from equation (II.2.1). Let the function $D(x', x)$ be defined by

$$D(x',x) = A(x',x) - A(x-x') \tag{2.0}$$

so that

$$D(x',-\infty) = D(x',\infty) = 0. \tag{2.0a}$$

The substitution $A(x',x) = D(x',x) + A(x-x')$ in (II.2.1) gives

$$F_k(x) = \int A(x-x')\,dF_{k-1}(x') + C_k(x) \quad \text{for } k \geqslant 1, \tag{2.1}$$

where $F_0(x)$ is prescribed, and for $k \geqslant 1$

$$C_k(x) = \int D(x',x)\,dF_{k-1}(x') = \int \{A(x',x) - A(x-x')\}\,dF_{k-1}(x'). \tag{2.2}$$

Every function $C_k(x)$ is therefore the difference between two distributions and has finite variation on $(-\infty,\infty)$. The equations (2.1) and (2.2) are valid for *all* x. The chain of equations (2.1) may now be employed to exhibit $F_k(x)$ in terms of the set of functions $\{C_k(x)\}$ and the Green distributions $G_k(x) = A^{(k)}(x)$ appearing in (II.3.3). If we define $C_0(x)$ to be $F_0(x)$ we have from (2.1) for all k

$$F_k(x) = \sum_{0}^{k} \int G_{k-k'}(x-x')\,dC_{k'}(x') \tag{2.3}$$

where

$$G_k(x) = A^{(k)}(x); \quad G_0(x) = A^{(0)}(x) = U(x). \tag{2.4}$$

If we now employ (2.3) with (2.2) to eliminate $F_k(x)$ we obtain a set of recursion equations generating the functions $C_k(x)$. Specifically, we find

$$C_0(x) = F_0(x) \tag{2.5a}$$

and

$$C_{k+1}(x) = \left[\sum_{0}^{k} C_{k'} \cdot G_{k-k'}\right] \cdot D \tag{2.5b}$$

where $A \cdot B$ denotes $\int B(x',x)\,dA(x')$. Since all operations are well defined and only finite sums are involved, the functions $C_k(x)$ are completely determined by (2.5). The representation (2.3) and its analogue in continuous time will underlie much of what follows. The function $C_k(x)$ acts as a correction term in (2.1), compensating for the spatial inhomogeneity of the single-step transition distribution $A(x',x)$. Because of this role the functions $C_k(x)$ will be called *compensation functions.*

Consider in particular the replacement process (1.5). Outside the interval $[0,L]$, where $A(x',x) = 0$ for $x < 0$ and $A(x',x) = 1$ for $x > L$, the compensation functions (2.2) reduce to the form

$$C_k(x) \;=\; -\int A(x-x')\,dF_{k-1}(x') \quad \text{for } x < 0, \tag{2.6a}$$

and

$$C_k(x) \;=\; \int \{1-A(x-x')\}\,dF_{k-1}(x') \quad \text{for } x > L. \tag{2.6b}$$

Consequently, we infer that for $k > 0$,

$$C_k(-\infty) \;=\; C_k(+\infty) \;=\; 0, \tag{2.7}$$

and that $C_k(x)$ will be localized about the boundaries outside $[0,L]$ to the extent to which the increments are small.

The structure and character of the compensation functions are more easily visualized for the replacement process (1.5) when $F_0(x)$, the increment distribution $A(x)$, and the replacement distributions $B_1(x)$ and $B_2(x)$ are absolutely continuous, and a local description is convenient. Let $a(x',x) = (d/dx)A(x',x)$ be the single-step transition density for transitions from x'. Then the local equivalent of (1.6) is

$$a(x',x) \;=\; a(x-x')[U(x)-U(x-L)] + A(-x')\,b_1(x) + \{1-A(L-x')\}\,b_2(x). \tag{2.8}$$

$F_k(x) = \int A(x',x)\,dF_{k-1}(x)$, and $F_k(x)$ will be absolutely continuous for $k \geqslant 1$, by virtue of the absolute continuity of $A(x',x)$. Moreover from (2.2), $C_k(x)$ will be absolutely continuous for all k. Consequently,

$$f_k(x) \;=\; \int_0^L f_{k-1}(x')\,a(x-x')\,dx' + c_k(x), \tag{2.9}$$

where $c_0(x) = f_0(x)$, and for $k > 0$,

$$c_k(x) = \int_0^L f_{k-1}(x')\{a(x', x) - a(x-x')\}\,dx'. \tag{2.10}$$

The analogue of (2.3) is

$$f_k(x) = \int g_k(x-x')\,dF_0(x') + \sum_1^k \int g_{k-k'}(x-x')\,c_{k'}(x')\,dx', \tag{2.11}$$

where $g_k(x) = a^{(k)}(x)$. The functions $c_j(x)$ in (2.11) appear as source densities and contribute to $f_k(x)$ via the Green densities $g_k(x)$. From (2.8) and (2.10), we have, for $k \geqslant 1$,

$$\begin{aligned} c_{k+1}(x) = & -\{1 - U(x)\} \int_0^L f_k(x')\,a(x-x')\,dx' \\ & - U(x-L) \int_0^L f_k(x')\,a(x-x')\,dx' \\ & + b_1(x) \left[\int_0^L f_k(x')\,A(-x')\,dx'\right] \\ & + b_2(x) \left[\int_0^L f_k(x')\{1 - A(L-x')\}\,dx'\right]. \end{aligned} \tag{2.12}$$

The sources consist, therefore, of four terms. The first term in (2.12) represents the neutralization of process samples that would go negative if the process were spatially homogeneous. The second term serves the same function for samples that would go beyond $x = L$. The third term represents the replacement of terms going negative with the replacement distribution $B_1(x)$, etc. Since for every sample neutralized there is one sample replacement, $\int_{-\infty}^{\infty} c_k(x)\,dx = 0$ and this is the meaning of (2.7). Similarly, equation (2.6) has a simple meaning from this point of view. The formal analysis does not require such interpretation but is motivated and facilitated thereby.

For the process (1.3) with retaining boundaries, the replacement distributions $U(x)$ and $U(x-L)$ are not absolutely continuous. It is easily seen, however, that if the increment distribution $A(x)$ is absolutely continuous, $F_k(x)$ will be absolutely continuous for $k \geqslant 1$ over every interval not containing the boundaries. Equation (2.11) is then replaced by

$$f_k(x) = \int g_k(x-x')\,dF_0(x') + \sum_1^k \int g_{k-k'}(x-x')\,dC_{k'}(x'), \qquad (2.13)$$

for $x \neq 0,L$.

V.3 Compensation functions for continuous time processes †

The nature of the compensation function as a source term for the Green's function is seen more vividly in connection with the associated processes in continuous time (Section II.2). If the transitions governed by $A(x',x)$ occur at Poisson epochs of frequency ν, the jump process $X(t)$ in continuous time has a distribution $F(x,t)$ satisfying the continuity equation (II.2.9) which reads

$$\frac{\partial F(x,t)}{\partial t} = -\nu F(x,t) + \nu \int A(x',x)\,dF(x',t). \qquad (3.0)$$

If we replace $A(x',x)$ by $A(x-x') + D(x',x)$ as in Section 2, we may then write (3.0) as

$$\frac{\partial F(x,t)}{\partial t} = -\nu F(x,t) + \nu \int A(x-x')\,dF(x',t) + C(x,t) \qquad (3.1)$$

where

$$C(x,t) = \nu \int D(x',x)\,dF(x',t). \qquad (3.2)$$

Equation (3.1) is the continuity equation (II.3.7) for the homogeneous step process of frequency ν and increment distribution $A(x)$, modified by the presence of the term $C(x,t)$ which may be regarded as a source term. A solution of (3.2) is provided by

$$F(x,t) = \int G(x-x',t)\,dF_0(x') + \int_0^t dt' \int G(x-x',t-t')\,dC(x',t') \qquad (3.3)$$

where $G(x,t)$ is the Green distribution (II.3.5) for the homogeneous step process. The structure of (3.3) is a direct analogue of that of (2.3). A

† A discussion of compensation for lattice processes which are not temporally homogeneous may be found in Keilson (1962a) and (1962b).

discussion of the uniqueness of the solution (3.3) may be avoided by the observation that (3.3) is a direct consequence of (2.3) and the relation

$$C(x,t) = \nu \sum_0^\infty \{e^{-\nu t}(\nu t)^k/k!\}\, C_{k+1}(x) \qquad (3.4)$$

obtained from (2.2), (3.2) and (II.2.5). The second term on the right-hand side of (3.3) has the form $C(x,t)\cdot * G(x,t)$ where the dot indicates convolution in x and the asterisk convolution in time.

When (3.4) and (II.3.5) are employed this term becomes

$$T_1(x,t) = \nu \sum_0^\infty \sum_0^\infty \{C_{j+1}(x)\cdot G_k(x)\}[\{e^{-\nu t}(\nu t)^j/j!\} * \{e^{-\nu t}(\nu t)^k/k!\}].$$

Via Laplace transformation, the time convolution in square brackets is found to be $\nu^{-1}\{e^{-\nu t}(\nu t)^{j+k+1}/(j+k+1)!\}$. From (2.3), however, $F(x,t) = \sum_0^\infty F_k(x)\{(\nu t)^k e^{-\nu t}/k!\} = F_0(x)\cdot G(x,t) + T_2(x,t)$, where

$$T_2(x,t) = \sum_{k=1}^\infty \{e^{-\nu t}(\nu t)^k/k!\} \sum_{k'=1}^{k} \{C_{k'}(x)\cdot G_{k-k'}(x)\}.$$

The double series $T_1(x,t)$ and $T_2(x,t)$ may now be seen to differ only in the order of terms. Since the series are absolutely convergent, rearrangement of terms leaves the series unchanged and $T_1(x,t) = T_2(x,t)$.

The function $C(x,t)$ is the compensation function for processes in continuous time, fully determined by $F(x,t)$ and (3.2). From (3.2), since $D(x',x) = A(x',x) - A(x-x')$, $C(x,t)$ is the difference between two distributions and hence has bounded variation over the interval $(-\infty,\infty)$. From (3.4) it may be seen that for $t > 0$, $C(x,t)$ has all order derivatives in time. Indeed the function $C(x,t)$ is an entire function of t. For in any circle $0 \leqslant |t| \leqslant T$, $|t^k/k!| \leqslant |T|^k/k!$ and $|C_k(x)| \leqslant 2$. Hence $|\sum_0^\infty \{t^k/k!\}C_{k+1}(x)| \leqslant 2e^T$, and the power series (3.4) defines an entire function. An equation governing $C(x,t)$ is obtained from (3.2) and has the form

$$C = \nu\{F_0\cdot G + C\cdot * G\}\cdot D. \qquad (3.5)$$

This equation will be examined subsequently. As for (2.8),

$$C(x,t) = -\nu \int A(x-x')\,dF(x',t) \qquad x < 0 \qquad (3.6)$$

and

$$C(x,t) \;=\; \nu \int [1 - A(x-x')]\, dF(x',t) \qquad x > L \tag{3.6}$$

so that

$$C(-\infty,t) \;=\; C(+\infty,t) \;=\; 0\,. \tag{3.7}$$

The homogeneous diffusion and drift processes discussed in Section 1 may be treated, as stated there, as the limit of a sequence of bounded homogeneous jump processes $X_j(t)$. The functions $F_j(x,t)$, $D_j(x',x)$, $C_j(x,t)$ and $G_j(x,t)$ for any $X_j(t)$ will satisfy equations (3.2) and (3.3). As $j \to \infty$ and the limiting processes are attained, their behaviour may be related to that for the jump processes. Caution is necessary, for the functions $C_j(x,t)$ and $D_j(x',x)$ will not have suitable limits in some cases of interest.

As an illustration of these remarks, consider the homogeneous drift process of drift rate $v < 0$, frequency ν_0 and positive increments with distribution $A_0(x)$. Let the process be modified by replacement at ζ with distribution $B(x)$ whenever the value $x = 0$ is reached (cf. Example 5a of Section 1). This process may be treated as the limit of a sequence of jump processes $X_j(t)$ with frequency ν_j and increment distribution $A_j(x)$ given by (1.7), and replacement distribution $B(x)$. For simplicity we assume $F_0(x)$, $A_0(x)$ and $B(x)$ are absolutely continuous. From (3.4) and (2.12) we then have for the compensation density

$$c_j(x,t) = |v| \int_0^{\infty} f_j(x',t)(\eta_j\, e^{-\eta_j x'})\, dx' \{-\eta_j\, e^{\eta_j x}\, U(-x) + b(x)\}\,. \tag{3.8}$$

The limiting drift process $X(t)$ corresponds to $j = \infty$, and $\eta_j = \infty$, and for this process $F(x,t)$ will still be absolutely continuous. From (3.8) it is easy to see that the compensation function $C_j(x,t)$ becomes

$$C(x,t) \;=\; |v|\, f(0+,t)\{-U(x) + B(x)\}\,. \tag{3.9}$$

Equation (3.3) will then be valid with $G(x,t)$ given by (II.3.9) and $C(x,t)$ by (3.9). For the drift process modified by a retaining boundary, where $B(x) = U(x)$, equation (3.9) loses all meaning.

The value of (3. 9) will be seen most clearly when we deal with an ergodic process whose stationary distribution is of interest (Section VI.5).

V.4 Bounded homogeneous processes on the lattice

The remarks of the previous three sections apply to lattice processes as a subcase. Because such lattice processes have a distinct character and notation, a brief discussion is appropriate.

The simplest bounded homogeneous lattice process is the analogue of (1.1) with

$$N_k = \max[0, N_{k-1} + \xi_k], \tag{4.1}$$

where N_0 is a lattice random variable of prescribed distribution $\sum p_n^{(0)} U(x-n)$ with g.f. $\pi^{(0)}(u)$ (Section II.4) and the increments ξ_k are independent lattice random variables having a common distribution with generating function $\epsilon(u) = \sum_{-\infty}^{\infty} \mathrm{e}_n u^n$. The single-step transition probabilities for (4.1) are, for $0 \leqslant n'$,

$$\begin{aligned} \mathrm{e}_{n'n} &= 0 & n < 0 \\ \mathrm{e}_{n'0} &= \sum_{-\infty}^{-n'} \mathrm{e}_j & n = 0 \\ \mathrm{e}_{n'n} &= \mathrm{e}_{n-n'} & n > 0 . \end{aligned} \tag{4.2}$$

The process (4.1) may be described as a spatially homogeneous lattice process modified by a single retaining boundary at $n = 0$. Similarly one may have processes with two retaining boundaries at $n = 0$ and $n = L$. The replacement processes of Section 1 have their lattice analogues with prescribed replacement lattice distributions $B_1(x) = \sum b_{1n} U(x-n)$, and $B_2(x) = \sum b_{2n} U(x-n)$. In place of (1.6) one would then have, for $0 \leqslant n' \leqslant L$, the single-step transition probabilities,

$$\begin{aligned} &\mathrm{e}_{n'n} = 0, \quad n < 0; \qquad\qquad \mathrm{e}_{n'n} = 0, \quad L < n; \\ &\mathrm{e}_{n'n} = \mathrm{e}_{n-n'} + b_{1n} \sum_{-\infty}^{-n'-1} \mathrm{e}_j + b_{2n} \sum_{L-n'+1}^{\infty} \mathrm{e}_j , \qquad 0 \leqslant n \leqslant L . \end{aligned} \tag{4.3}$$

For the process with two retaining boundaries at $n = 0$ and $n = L$, $b_{1n} = \delta_{n,0}$ and $b_{2n} = \delta_{n,L}$.

The compensation functions of Section 2 for the processes in discrete time are obtained by introducing the analogue of (2.0), defined for $0 \leqslant n' \leqslant L$ by

$$d_{n'n} = \mathrm{e}_{n'n} - \mathrm{e}_{n-n'} \tag{4.4}$$

We then have, as for (2.1) and (2.2), for *all* values of n and $k \geqslant 1$,

$$p_n^{(k)} = \sum_0^L p_{n'}^{(k-1)} e_{n-n'} + c_n^{(k)} \tag{4.5}$$

where

$$c_n^{(k)} = \sum_0^L p_{n'}^{(k-1)} d_{n'n} = \sum_0^L p_{n'}^{(k-1)} (e_{n'n} - e_{n-n'}). \tag{4.6}$$

Hence (2.3) takes the form, when $c_n^{(0)}$ is defined to be $p_n^{(0)}$,

$$p_n^{(k)} = \sum_{-\infty}^{\infty} \sum_{k'=0}^{k} c_{n'}^{(k')} g_{n-n'}^{(k-k')} \tag{4.7}$$

where $g_n^{(k)} = e_n^{(k)}$ are the Green probabilities of Section II.4.

From (4.6) we see that $c_n^{(k)}$ is the difference between two sets of probabilities, so that

$$\sum_{n=-\infty}^{+\infty} |c_n^{(k)}| \leqslant 2, \tag{4.8}$$

and $\lim_{|n|\to\infty} c_n^{(k)} = 0$.

For continuous time, (3.2) and (3.4) become, for all n,

$$c_n(t) = \nu \sum_0^{\infty} \{e^{-\nu t}(\nu t)^k/k!\}\, c_n^{(k+1)} \tag{4.9}$$

and

$$c_n(t) = \nu \sum_0^L p_{n'}(t)\, d_{n'n} = \nu \sum_0^L p_{n'}(t)(e_{n'n} - e_{n-n'}). \tag{4.10}$$

Hence $c_n(t)$ is the difference between two probabilities and $\sum_{-\infty}^{\infty} |c_n(t)| \leqslant 2$. The lattice analogue of (3.3) for continuous time is

$$p_n(t) = \sum_{n'} p_{n'}^{(0)} g_{n-n'}(t) + \sum_{n'} c_{n'}(t) * g_{n-n'}(t) \tag{4.11}$$

where the asterisk denotes convolution in time.

V.5 The role of the extended renewal function as ergodic Green measure

The remaining sections of this chapter will be devoted to processes which settle down to a unique stationary distribution, given sufficient time, from any prescribed initial state.

A temporally homogeneous Markov process in discrete or continuous time will be called *ergodic* if for any initial distribution $F_0(x)$, the process has a limiting† distribution $F_\infty(x)$, as k or $t \to \infty$, independent of $F_0(x)$ and of finite non-zero variance. The term ergodic, as normally employed in the literature, has reference to processes which at $t = 0$ are already in their stationary distributions. Rather than devise a new term, we will use ergodic for processes with prescribed initial distributions as well. The unique stationary distribution for the ergodic process will be called the *ergodic distribution.*

For the process X_k in discrete time governed by the single-step transition distribution $A(x', x)$, ergodicity will imply that $\lim_{k\to\infty} F_k(x) = F_\infty(x)$. For the associated class of processes in continuous time with the same $A(x',x)$ and Poisson transition epochs of frequency ν (Section II.1), it is easy to prove that

$$\lim_{t\to\infty} F(x,t) = \lim_{t\to\infty} \sum_0^\infty \{e^{-\nu t}(\nu t)^k/k!\} F_k(x) = \lim_{k\to\infty} F_k(x) = F_\infty(x) \tag{5.1}$$

i.e., ergodicity of the discrete-time process implies ergodicity for the associated processes in continuous time.

From equation (2.2), when X_k is ergodic, the compensation functions $C_k(x)$ have a limit, when $k \to \infty$, given by

$$C_\infty(x) = \int D(x',x)\,dF_\infty(x') = \int \{A(x',x) - A(x-x')\}\,dF_\infty(x'). \tag{5.2}$$

By virtue of (5.2), $C_\infty(x)$ is the difference between two distribution functions. The ergodic distribution is obtained by passing to the limit $k = \infty$ in (2.3). For clarity consider the replacement process (1.5) where $A(x)$ and the replacement distributions $B_1(x)$ and $B_2(x)$ are absolutely continuous, so that (2.11) and (2.12) are valid. It is then

---

† The notation $\lim_{k\to\infty} F_k(x) = F_\infty(x)$ will imply that the entity $F_k(x)$ approaches $F_\infty(x)$ in a suitable topology. (See e.g., Lukacs, 1960, Section 3.4.)

clear from (5.2) that $C_\infty(x)$ will be absolutely continuous. The first term on the right of (2.11) goes to zero as $x \to \infty$, since $\lim_{k\to\infty} g_k(x) = 0$ for every finite x. For the second term on the right of (2.11) we have symbolically $\lim_{k\to\infty} (g_0 \cdot c_k + g_1 \cdot c_{k-1} \cdots\cdots)$ and one expects the limiting form $\{\sum_0^\infty g_i(x)\} \cdot c_\infty(x)$. Since $g_k(x) = a^{(k)}(x)$,

$$f_\infty(x) = c_\infty(x) + \int h(x-x')\, c_\infty(x')\, dx' \tag{5.3}$$

where

$$h(x) = \sum_1^\infty a^{(n)}(x), \tag{5.4}$$

provided that the integral in (5.3) and the series (5.4) converge.

The function $h(x)$ defined by (5.4) is known in renewal theory as the *extended renewal density*. When all increments are positive, i.e., when $A(0) = 0$, $h(x)$ is the ordinary renewal density, and the existence of a first moment for $A(x)$ is known to assure the convergence of (5.4). For absolutely continuous distributions $A(x)$ with both positive and negative support, a finite first moment for $A(x)$ *not equal to zero* assures the convergence† of (5.4). A demonstration of this will be given under slightly restrictive conditions on $A(x)$ in Section 6. That the limit (5.3) is approached by $f_k(x)$ as $k \to \infty$, will be shown in Section 11.

A simple grasp of the importance of the requirement that the first moment must not vanish may be obtained from central limit considerations. Consider a density $a(x)$, bounded on $(-\infty,\infty)$, having a finite second moment and zero mean. For fixed x, $a^{(k)}(x) \sim (2\pi k\sigma^2)^{-\frac{1}{2}}$ as $k \to \infty$ (see the proof of Theorem III.5.1 of Note 4), and (5.4) for such a density diverges.

Equation (5.3) may be written in the form

$$f_\infty(x) = \int c_\infty(x')\, g(x-x')\, dx' \tag{5.5}$$

where $g(x)$ is the generalized function

---

† Weaker conditions for the convergence of (5.4) and (5.7) are given by Feller and Orey (1961) and Smith (1962). An excellent review of renewal theory and a history of earlier studies of the convergence of (5.4) may be found in Smith (1958).

$$g(x) = \delta(x) + h(x). \tag{5.6}$$

The function $g(x)$ will be called the *ergodic Green density.* From (5.5) we see that $g(x)$ has the character of a Green's function for the ergodic density $f_\infty(x)$ of X_k with the compensation function $c_\infty(x)$ serving as the source density.

Now let $H(x)$ be the *extended renewal function* defined by

$$H(x) = \sum_{1}^{\infty} \{A^{(k)}(x) - A^{(k)}(0)\}. \tag{5.7}$$

When the extended renewal density $h(x)$ of (5.4) exists, one will have $H(x) = \int_0^x h(y)\,dy$.† Since any distribution may be represented as the limit of a sequence of absolutely continuous distributions, one anticipates that the series (5.7) will be convergent when the distribution $A(x)$ has a non-zero first moment. A demonstration of this convergence, when a convergence strip of finite width is available for $\alpha(z)$, will be given in Section 6.1.

The extended renewal function $H(x)$ has the following probabilistic interpretation. Consider an ensemble of samples $\{x_1(\omega), x_2(\omega) \ldots.\}$ for the process $X_k = \sum_1^k \xi_i$ where the ξ_i are independent and have distribution $A(x)$. For any sample ω let $n_\omega(y_1, y_2)$ be the number of values $x_k(\omega)$ falling in $(y_1, y_2]$. The expectation of $N(y_1, y_2)$ is $H(y_2) - H(y_1)$. If $\mu = E(\xi) \neq 0$, the convergence of the extended renewal function implies that a finite interval will only be visited a finite number of times. The expectation of the number of visits to a set $S_1 + S_2$ on $(-\infty, \infty)$ is the sum of the expectations for S_1 and S_2 when S_1 and S_2 are disjoint. The expectation is therefore an additive set function and $H(x)$ is its *measure function.*

An integration of (5.3) from 0 to x gives $F_\infty(x) - F_\infty(0) = [C_\infty(x) + H(x) \cdot C_\infty(x)]_0^x$ where $[K(x)]_a^b$ denotes $K(b) - K(a)$. We thus anticipate for the integral form of (5.5), the relation‡

$$F_\infty(x) - F_\infty(0) = [C_\infty(x) \cdot G(x)]_0^x \tag{5.8}$$

† This is justified by the positive character of $a^{(n)}(x)$ which permits the interchange of summation and integration when either operation converges. We see that the absolute continuity of $A(x)$ implies the absolute continuity of $H(x)$ in every finite interval. See Note 7.

‡ See first footnote on next page.

where $\mathbf{G}(x)$ is defined by

$$\mathbf{G}(x) = U(x) + H(x). \tag{5.9}$$

A more careful discussion of the validity of (5.8) as a limiting form will be found in Section 11.

The function $\mathbf{G}(x)$ will be called the *ergodic Green measure* §. From (5.7) and (5.9) we see that $\mathbf{G}(x)$ is monotonic increasing and positive for $x \geqslant 0$. It will be seen subsequently that $H(x)$ and $\mathbf{G}(x)$ are unbounded.

In the sections to follow the importance of (5.5) and (5.8) as a key to the structure and asymptotic behaviour of the ergodic distribution of X_k will be demonstrated.

The basic character of equations (5.2) and (5.3) is maintained for the ergodic jump processes in continuous time. A comparison of (3.2) and (5.2) gives

$$C(x,\infty) = \lim_{t\to\infty} C(x,t) = \nu\, C_\infty(x). \tag{5.10}$$

Examination of (3.3) shows that the first term on the right goes to zero for any fixed x, and in the presence of ergodicity, as for (5.8),

$$\lim_{t\to\infty} \{F(x,t) - F(0,t)\} = \left[C(x,\infty) \cdot \int_0^\infty \{G(x,t) - G(0,t)\}\, dt \right]_0^x \tag{5.11}$$

provided that the infinite integral is meaningful. From (II.4.5) however, we have

---

‡ See previous page.
The function $H^+(x) = \sum_1^\infty A^{(k)}(x)$ converges when $A(x)$ has a positive first moment and $V_{\max}$, the upper boundary of the convergence strip of $\alpha(z)$, is positive (Section 6.1). Equation (5.8) may then be replaced by the more convenient relation

$$F_\infty(x) = C_\infty(x) + C_\infty(x) \cdot H^+(x). \tag{5.8$'$}$$

Similarly when $A(x)$ has a negative first moment and $V_{\min} < 0$, $H^-(x) = \sum_1^\infty \{A^{(k)}(x) - 1\}$ converges and (5.8) may be replaced by

$$F_\infty(x) = 1 + C_\infty(x) + C_\infty(x) \cdot H^-(x). \tag{5.8$''$}$$

§ The measure functions employed in this book are called simply measures for convenience.

$$\int_0^\infty \{G(x,t) - G(0,t)\,dt = \nu^{-1}\sum_0^\infty \{A^{(j)}(x) - A^{(j)}(0)\} = \nu^{-1}[H(x) + U(x)]_0^x. \tag{5.12}$$

Equation (5.11) may therefore be written as

$$F(x,\infty) - F(0,\infty) = [C(x,\infty)\cdot \mathbf{G}_c(x)]_0^x \tag{5.13}$$

where

$$\mathbf{G}_c(x) = \int_0^\infty \{G(x,t) - G(0,t)\,dt = \nu^{-1}\,\mathbf{G}(x). \tag{5.14}$$

The function $\mathbf{G}_c(x)$ will be called the *ergodic Green measure for continuous time*. By virtue of (5.14) and (5.10) the ergodic distributions obtained from (5.8) and (5.11) are in agreement.

For the diffusion processes and drift processes of Section 1 and Section II.7 characterized as the limit of a sequence of step processes $X_j(t)$, the Green measure $\mathbf{G}_c(x)$ defined by the integral in (5.14) will be convergent and meaningful, provided that $G(x,1)$ has a finite non-zero mean, i.e.,

$$\int x\,dG(x,1) \neq 0. \tag{5.15}$$

Consider for example the diffusion process with time-dependent Green density $g_D(x,t)$ given by (II.3.23). The ergodic Green measure (5.14) is then absolutely continuous, and has a local density

$$\begin{aligned} \mathbf{g}_c(x) &= \int_0^\infty g_D(x,t)\,dt \\ &= \int_0^\infty \exp\{-(4Dt)^{-1}(x-vt)^2\}\{4\pi Dt\}^{-\frac{1}{2}}\,dt \\ &= |v|^{-1}\exp\left\{\frac{vx - |vx|}{2D}\right\} \quad \text{[Magnus and Oberhettinger, p. 128].} \end{aligned}$$

Note that when $v = 0$, the ergodic Green density does not converge. Note also that for $vx > 0$, $\mathbf{g}_c(x) = |v|^{-1}$ and is constant. This structural behaviour of $\mathbf{g}(x)$ will be found subsequently under certain more general conditions, and will be of considerable importance.

(For this diffusion process the local form of (3.1) is $\frac{\partial f}{\partial t} = D\frac{\partial^2 f}{\partial x^2} - v\frac{\partial f}{\partial x} + c(x,t)$. If as $t \to \infty$, $c(x,t) \to c(x,\infty)$, $f(x,\infty)$ would obey the equation for all x

$$D\frac{\partial^2}{\partial x^2} f(x,\infty) - v\frac{\partial}{\partial x} f(x,\infty) \;=\; -c(x,\infty). \tag{5.16}$$

The ergodic Green density is a solution of (5.16) for $c(x,\infty) = \delta(x)$. A derivation of the ergodic Green density via the differential equation is undesirable because of the difficulty in establishing boundary conditions at infinity. For diffusion processes on a finite interval, the boundary conditions are more readily available and the differential equation approach to the ergodic distribution is effective.)

The validity of (5.15) for the homogeneous drift process will be demonstrated in Section 9. The validity of (5.15) for more general homogeneous processes in continuous time will be discussed in Section 9.1.

V.6 The extended renewal density

When $A(x)$ is absolutely continuous, the derivative of the extended renewal function (5.7) is the extended renewal density,

$$\frac{d}{dx} H(x) \;=\; h(x) \;=\; \sum_{1}^{\infty} a^{(k)}(x). \tag{6.1}$$

We will demonstrate the convergence of the series (6.1) when $a(x)$ has a non-zero first moment and is reasonably well behaved, as specified in the following two theorems† due in the main to W. L. Smith (1955b, 1962).

Theorem 6.1. If (1) the distribution $A(x)$ is absolutely continuous and has a finite non-zero mean μ; (2) its density $a(x)$ is bounded for all x; (3) $x\,a(x)\cdot\log(1+|x|) \in L_1(-\infty,\infty)$; then

(a) the series $h(x) - a(x) = \sum_2^\infty a^{(k)}(x)$ is convergent, bounded, and continuous for all x;

(b) if $\mu > 0$, $h(x) - a(x)$ goes to 0 at $x = -\infty$, and goes to $|\mu|^{-1}$ at $x = +\infty$. If $\mu < 0$ the behaviour at $+\infty$ and $-\infty$ is reversed. If $\mu = 0$, the series diverges.

To prove this theorem, we first show its truth when the characteristic function $\alpha(z)$ is rational. The absolute continuity of $A(x)$

† It is shown in Note 8 that (3) may be replaced by (3′): $x\,a(x) \in L_1$.

implies that $\alpha(z)$ is of degree $\leqslant -1$. Consider the function

$$h(w,x) = \sum_{1}^{\infty} w^n a^{(n)}(x) \tag{6.2}$$

where $0 \leqslant |w| < 1$. The function $|h(w,x)|$ is integrable and hence $h(w,x)$ has a Fourier transform given by the series $\sum_{1}^{\infty} w^n \alpha^n(z) = w\,\alpha(z)/\{1 - w\,\alpha(z)\}$. The inversion

$$h(w,x) = w\,a(x) + \frac{1}{2\pi}\int_{-\infty}^{+\infty} \frac{w^2 \alpha^2(z)}{1 - w\,\alpha(z)}\, e^{-izx}\, dz \tag{6.3}$$

is valid since $\alpha(z) = 0(|z|^{-1})$ at infinity. The function $w^2\alpha^2(z)/\{1 - w\,\alpha(z)\}$ is rational with singularities of $\alpha(z)$ and at the zeros of the denominator. When $\mu \neq 0$, $\alpha(iV)$ will have a minimum (cf. Section III.2) $\alpha(iV_0)$ at some value V_0, positive for $\mu > 0$, and negative for $\mu < 0$, in the convergence strip. The contour of integration of (6.3) may be translated to the line $z = U + iV_0$ since $|1 - w\,\alpha(z)| > 0$ in the intervening strip, and (6.3) becomes

$$h(w,x) = w\,a(x) + \frac{1}{2\pi}\int_{-\infty+iV_0}^{\infty+iV_0} \frac{w^2 \alpha^2(z)}{1 - w\,\alpha(z)}\, e^{-izx}\, dz\,. \tag{6.4}$$

Since $\alpha(iV_0) < 1$, it is clear from (6.4) that $h(w,x)$ is analytic in w at $w = 1$. Consequently $h(x) = h(1,x)$ and the convergence stated in (a) is proved.

The function $h(1,x)$ may be evaluated from (6.4) by residues for $x > 0$ if the contour is closed in the lower half-plane at infinity. For $x < 0$, the contour is closed in the upper half-plane at infinity. Since $\mu \neq 0$, $[1 - \alpha(z)]^{-1}$ has a simple pole at $z = 0$, with residue $i\mu^{-1}$. All other zeros give rise to exponentially decaying contributions to $h(x)$. The function $h(x)$ may be written as

$$h(x) = a(x) + \frac{1}{2\pi}\, e^{V_0 x} \int_{-\infty}^{\infty} \frac{\alpha^2(iV_0 + U)}{1 - \alpha(iV_0 + U)}\, e^{-iUx}\, dU\,, \tag{6.5}$$

for which, since $\alpha(z)$ is rational and of degree $\leqslant -1$, the integrand is absolutely integrable. The stated continuity of $h(x)$ then follows. Since, moreover, $h(x)$ is clearly bounded in every finite interval, and asymptotically zero or $|\mu|^{-1}$, $h(x)$ is bounded.

When $\mu = 0$, $V_0 = 0$ and the contour is not translated. The function $h(w,x)$ may again be evaluated by residues. As $w \to 1$, there will be one zero of $1 - w\alpha(z)$ in each half-plane approaching $z = 0$, and one finds that $\Sigma w^n a^{(n)}(x)$ goes to $+\infty$ for all x.

Consider now the more general $a(x)$. Let $b(x)$ be any second density of the type just considered having a rational characteristic function. Then we may write

$$\mathfrak{F}\{h(w,x)\} = \frac{w\,\alpha(z)}{1 - w\,\alpha(z)}$$

$$= \left\{1 + \frac{w\,\beta(z)}{1 - w\,\beta(z)}\right\} \cdot \left\{\frac{w^2\alpha(z)[\alpha(z) - \beta(z)]}{1 - w\,\alpha(z)} + w\,\alpha(z)\right\}. \quad (6.6)$$

The behaviour of the inverse transform as $w \to 1$ may be deduced from that of

$$h_b(w,x) = \sum_1^\infty w^n b^{(n)}(x),$$

and

$$j(w,x) = \mathfrak{F}^{-1}\left\{\frac{w^2\alpha(z)[\alpha(z) - \beta(z)]}{1 - w\,\alpha(z)} \quad \frac{1}{1 - w\,\beta(z)}\right\}.$$

We ask that $b(x)$ be chosen such that $\mu_b = \mu_a$, i.e. $\alpha'(0) = \beta'(0)$. Since $a(x)$ is bounded, $\alpha(z) \in L_2$ and $|\alpha(z)\beta(z)| \leqslant \frac{1}{2}\{|\alpha(z)|^2 + |\beta(z)|^2\} \in L_1$. The Fourier transform of $j(w,x)$ is then† seen to be absolutely integrable for all $0 \leqslant w \leqslant 1$. Hence for all such w, $j(w,x)$ is bounded and continuous in x, and by the Riemann-Lebesgue lemma, $j(w,x) \to 0$ as $x \to \pm\infty$. We may now write from (6.6), $h(w,x) = w h_b(w,x) \cdot a(x) + w a(x) + j(w,x)$ and pass to the limit $w = 1$. Since $z^{-2}\{\alpha(z) - \beta(z)\} \in L_1(-\delta, \delta)$ for $\delta > 0$, and $|\{1 - w\alpha(z)\}\{1 - w\,\beta(z)\}|^{-1} < 4|\{1 - \alpha(z)\}\{1 - \beta(z)\}|^{-1}$ for real z‡ it follows from dominated convergence that $j(w,x) \to j(1,x)$ as $w \to 1$. Moreover $h_b(w,x) \cdot a(x) \to h_b(x) \cdot a(x)$ and we have

$$h(x) = \lim_{w \to 1} h(w,x) = h_b(x) \cdot a(x) + a(x) + j(1,x). \quad (6.7)$$

† When $a(x)$ has a second moment, this is clear. See Note 8.

‡ For any w, ζ with $0 \leqslant w \leqslant 1$, $|\zeta| < 1$, we have $|(1 - w\zeta)^{-1} - (1 - \zeta)^{-1}| = |\zeta(1 - \zeta)^{-1}||(1 - w)(1 - w\zeta)^{-1}| < |1 - \zeta|^{-1}$ since $|(1 - w)(1 - w\zeta)^{-1}| < 1$. Hence $|(1 - w\zeta)^{-1}| < 2|1 - \zeta|^{-1}$.

From the asymptotic behaviour of $h_b(x)$, we then infer† that $h(x)$ has the behaviour stated in the theorem, completing the proof of Theorem 6.1.

Densities $a(x)$ having singularities at one or more values of x may often be discussed with the help of the following extension of Theorem 6.1.

Theorem 6.2. If $A(x)$ satisfies conditions (1) and (3) of Theorem 6.1 and for some value N, $a^{(N)}(x)$ is continuous and bounded for all x, then the series $\sum_{N+1}^{\infty} a^{(k)}(x) = h(x) - a(x) \ldots -a^{(N)}(x)$

(a) is convergent, bounded, and continuous for all x;

(b) has the asymptotic behaviour in (b) of Theorem 6.1.

For a rearrangement of terms gives

$$h(x) = \{a(x) + a^{(2)}(x) \ldots + a^{(N)}(x)\} \cdot \{\delta(x) + h_N(x)\} \qquad (6.8)$$

where $h_N(x)$ is the extended renewal density (6.1) for $a(x) = a^{(N)}(x)$. By our assumptions‡ and Theorem 6.1 the extended renewal density $h_N(x)$ is bounded and continuous for all x and satisfies (b) of Theorem 6.1 with $\mu_N = N\mu$. The convergence of the convolution (6.8) is then apparent. Statement (b) of the theorem may then be verified with the aid of (6.8) and Lemma 6.1$'$.

The simplest application of Theorem 6.2 is to the Gamma Distribution with density $a(x) = \{\Gamma(\kappa)\}^{-1} x^{\kappa-1} e^{-x} U(x)$, $0 < \kappa < 1$, which is finite and continuous for $x \neq 0$. Since a value N for which $N\kappa > 1$ may be found, and $a^{(N)}(x) = \{\Gamma(N\kappa)\}^{-1} \cdot x^{N\kappa-1} e^{-x} U(x)$ is bounded and continuous for all x, $h(x)$ is finite and continuous for $x \neq 0$.

A closed expression may be given for the extended renewal density in simple cases by integration in the complex plane.

Example A: Exponentially distributed increments

When $a(x) = q\eta e^{\eta x} U(-x) + p\lambda e^{-\lambda x} U(x)$, where $p+q = 1$, one finds

---

† We have made tacit use of the following lemma, which is easily verified. *Lemma V.6.1$'$.* If a function $h(x)$ is bounded and if $\lim_{x\to\infty} h(x) = h_\infty$, then for any distribution $A(x)$, $\lim_{x\to\infty} \int_{-\infty}^{\infty} h(x-x')\,dA(x') = h_\infty$.

‡ This is again clear when $a(x)$ has a second moment. See Note 8.

that $\mu = p\lambda^{-1} - q\eta^{-1}$ and $\alpha(z) = q\eta(\eta + iz)^{-1} + p\lambda(\lambda - iz)^{-1}$. Suppose $\mu > 0$. Then $h(x)$ may be evaluated from (6.4) with $w = 1$. For $x > 0$, one closes the contour in the upper half-plane and obtains

$$h(x) = \mu^{-1}, \qquad x > 0. \tag{6.9a}$$

By closing in the lower half-plane, one finds

$$h(x) = \frac{pq(\lambda + \eta)^2}{p\eta - q\lambda} \exp\{(p\eta - q\lambda)\, x\}, \qquad x < 0. \tag{6.9b}$$

The same result may be obtained from $\alpha(z)/\{1 - \alpha(z)\}$ by a partial fraction decomposition.

V.6.1 Convergence properties of the extended renewal function

Consider the function

$$H^{+}(w,x) = \sum_{1}^{\infty} w^n A^{(n)}(x) \tag{6.10}$$

where $A(x)$ is an increment distribution whose characteristic function $\alpha(z)$ has a convergence strip with boundaries $V_{\min}$ and $V_{\max}$. For $0 \leqslant |w| < 1$, the convergence of (6.10) for all x is apparent, since $A^{(n)}(x) \leqslant 1$. Let $A_V(x)$ be the conjugate distribution (Section III.2) associated with the value $z = iV$ in the convergence strip. Formally we may write from $A(x) = \int_{-\infty}^{x} dA(x')$, (III.2.1), and (6.10),

$$H^{+}(w,x) = \int_{-\infty}^{x} e^{Vx'}\, d\left[\sum_{1}^{\infty} w^n \alpha^n(iV)\, A_V^{(n)}(x')\right]. \tag{6.11}$$

The function in square brackets will have a finite total variation on $(-\infty,\infty)$ and the integral (6.11) will be absolutely convergent, provided that $V > 0$ and $|w|\,\alpha(iV) < 1$. Similar conclusions may be drawn for

$$H^{-}(w,x) = \sum_{1}^{\infty} w^n \{A^{(n)}(x) - 1\}$$

$$= -\int_{x}^{\infty} e^{Vx'}\, d\left[\sum_{1}^{\infty} w^n\, \alpha^n(iV)\, A_V^{(n)}(x')\right]. \tag{6.12}$$

The series converges if $|w| < 1$, and the integral converges if $V < 0$

and $|w|\alpha(iV) < 1$. It is clear from (6.11) that the power series $H^+(w,x) = \sum_1^\infty w^n A^{(n)}(x)$ has a radius of convergence ρ_+ for which

$$\rho_+ \geqslant [\alpha(iV_+^*)]^{-1} \tag{6.13}$$

where V_+^* is the non-negative value of V in the convergence strip at which $\alpha(iV)$ has its smallest value. When $V_{max} = 0$, $V_+^* = 0$ and $\rho_+ = 1$. For the power series $H^-(w,x)$ we find in the same way that

$$\rho_- \geqslant [\alpha(iV_-^*)]^{-1} \tag{6.14}$$

where V_-^* is the non-positive value of V in the convergence strip at which $\alpha(iV)$ has its minimum.

Clearly, the presence of a first moment is not necessary. A radius of convergence for $H^+(w,x)$ or $H^-(w,x)$ greater than unity is available when the convergence of the characteristic function $\alpha(z)$ extends into a half-plane in which $\alpha(iV)$ is less than 1. The situation is summarized in the following theorem.†

Theorem V.6.3. For any increment distribution $A(x)$, let $\alpha(iV_+^*)$ be the greatest lower bound of the set of values $\{\alpha(iV) : 0 \leqslant V < V_{max}\}$, in the convergence strip; let $\alpha(iV_-^*)$ be the greatest lower bound of $\{\alpha(iV) : V_{min} < V \leqslant 0\}$. Then the series $H^+(w,x) = \sum_1^\infty w^n A^{(n)}(x)$ has a radius of convergence $\rho_+ \geqslant [\alpha(iV_+^*)]^{-1}$; the series $H^-(w,x) = \sum_1^\infty w^n\{A^{(n)}(x) - 1\}$ has a radius of convergence $\rho_- \geqslant [\alpha(iV_-^*)]^{-1}$.

On the basis of this theorem the convergence of $H^+(x) = \sum_1^\infty A^{(n)}(x)$ and that of $H^-(x) = \sum_1^\infty \{A^{(n)}(x) - 1\}$ may be discussed.

Corollary V.6.3. If $V_{max} > 0$, and $\mu = \int x\, dA(x) > 0$, then the series $H^+(x) = \sum_1^\infty A^{(n)}(x)$ converges, and $\sum_1^\infty n^K A^{(n)}(x)$ converges for all positive K. Similarly if $V_{min} < 0$ and $\mu = \int x\, dA(x) < 0$, the series

---

† When $\alpha(iV)$ assumes a local minimum at $V = V_0 > 0$ in the convergence interval with a zero left-derivative at V_0, the conjugate distribution $A_{V_0}(x)$ has zero mean, and the series $H_{V_0}^+(x) = \sum_1^\infty A_{V_0}^{(k)}(x)$ diverges for all x as is easily shown. It follows from (6.11) that the radius of convergence of (6.10) is given exactly by $\rho_+ = [\alpha(iV_0)]^{-1}$, and that the series (6.10) diverges for $w = \rho_+$. Similarly, if $\alpha(iV)$ has a local minimum at $V_0 < 0$ with a zero right-derivative, $\rho_- = [\alpha(iV_0)]^{-1}$ and there is divergence at the radius of convergence itself.

$H^-(x) = \sum_1^\infty \{A^{(n)}(x) - 1\}$ converges, as also does $\sum_1^\infty n^K\{A^{(n)}(x) - 1\}$ for all positive K.

For $\rho_+ > 1$ and $\rho_- > 1$ in the respective cases, and the stated convergence follows from the regularity of $H^+(w,x)$ and $H^-(w,x)$ inside their circles of convergence.

From Theorem 6.3 we have at once the following important consequence.

Theorem V.6.4. Let $A(x)$ be the distribution of a random variable with purely positive support. Then the series

$$H^+(w,x) = \sum_1^\infty w^n A^{(n)}(x) \tag{6.15}$$

is an entire function of w (i.e., $\rho_+ = \infty$), for every value of x.

As proof we need only note that such a distribution has a convergence strip which always contains the half-plane $V > 0$, and that $\alpha(iV) \to 0$ as $V \to \infty$, so that $\rho_+ = \infty$ from Theorem 6.3. (See Note 12.)

Note that no assumptions of any kind have been made about the moments of $A(x)$. The standard convergence theorem of renewal theory stating that (6.15) converges for $w = 1$ when $A(x)$ has a positive first moment, is a corollary. The convergence of the series (6.15) for all w is a consequence of the Volterra character of the series. By this we mean that when $A(x)$ is absolutely continuous, $h(w,x) = \sum_1^\infty w^n a^{(n)}(x)$ is the unique solution of the Volterra integral equation

$$h(x) = w\,a(x) + w\int_0^x a(x-x')h(x')dx'. \tag{6.16}$$

The uniqueness of the solution and its entire character in w are familiar results in Volterra theory.

The radius of convergence of $H(w,x) = \sum_1^\infty w^n\{A^{(n)}(x) - A^{(n)}(0)\}$ may now be discussed. It is readily seen that: (A) when $H^+(w,x)$ converges, $H(w,x) = H^+(w,x) - H^+(w,0)$; (B) when $H^-(w,x)$ converges, $H(w,x) = H^-(w,x) - H^-(w,0)$; (C) the radius of convergence ρ of $H(w,x)$ is $\rho = \max\{\rho_-,\rho_+\}$.†

† The convergence of $H(x)$ implies that of $h(x)$ almost everywhere, when $A(x)$ is absolutely continuous (Note 7).

When the convergence strip of $\alpha(z)$ has zero width, the existence of a mean ($\neq 0$) for $A(x)$ assures the convergence of $H(x) = \sum_1^\infty \{A^{(n)}(x) - A^{(n)}(0)\}$ for any finite x. (See Note 8.) It is also known (Blackwell, 1953) that, for any $A(x)$ with mean $\mu > 0$ which does not have all support on a lattice,

$$\lim_{x \to +\infty} \{H(x+h) - H(x)\} = \mu^{-1} h. \tag{6.17}$$

When $\mu < 0$, $\lim_{x \to -\infty} \{H(x+h) - H(x)\}$ is of course $|\mu^{-1}| h$.

V.7 The ergodic Green's function for lattice processes

We next consider the lattice processes. When the process N_k is ergodic, the probability $p_n^{(k)}$ will approach some limiting probability

$$p_n^{(\infty)} = \lim_{k \to \infty} p_n^{(k)} \tag{7.1}$$

independent of $\{p_j^{(0)}\}$. The associated processes $N(t)$ in continuous time will also be ergodic and have the same ergodic distribution. For the compensation components (4.6) we then have the corresponding limit

$$c_n^{(\infty)} = \sum_0^L p_{n'}^{(\infty)} d_{n'n} = \sum_0^L p_{n'}^{(\infty)} (e_{n'n} - e_{n-n'}). \tag{7.2}$$

From (4.5) we may derive as above the lattice analogue of (5.5), given by

$$p_n^{(\infty)} = c_n^{(\infty)} + \sum c_{n'}^{(\infty)} h_{n-n'} = \sum c_{n'}^{(\infty)} g_{n-n'}, \tag{7.3}$$

where

$$h_n = \sum_{k=1}^{\infty} e_n^{(k)} \tag{7.4}$$

and

$$g_n = \delta_{n,0} + h_n. \tag{7.5}$$

Equation (7.3) is predicated on the convergence of the series (7.4), a discussion of which will soon be given. The ergodic Green components g_n of (7.5) are the lattice analogue of the ergodic Green density $g(x) = \delta(x) + h(x)$ appearing in (5.6). When increments are positive, the numbers h_n have a simple renewal interpretation. If e_n is the probability that an event at $t = 0$ recurs for the first time at $t = n$, then h_n

is the probability of any recurrence at $t = n$. For increments which may be positive or negative, the probabilistic interpretation of Section 5 is appropriate. The number h_n is the expectation of the number of visits paid to the state n by the homogeneous lattice process N_k with increment probabilities e_n, commencing at $N_0 = 0$. The extended renewal numbers h_n may then be greater than one. We now examine the convergence of the series (7.4) defining these numbers. Theorem 6.1 has the following analogue for the lattice.

Theorem V.7.1. Let N_k be a homogeneous lattice process $N_k = \sum_1^k \xi_i$, governed by increment probabilities $\{e_n\}$. (A) If $\Sigma|n| \log(1+|n|) e_n < \infty$ and $\mu = \Sigma n\, e_n \neq 0$, the series (7.4) converges. (B) If $\mu = 0$, the series diverges. (C) If, moreover, the magnitudes of possible increments, i.e. $\{|n| \,\big|\, e_n > 0\}$, have unity as their greatest common divisor, and $\mu > 0$, then $\lim_{n\to\infty} h_n = |\mu|^{-1}$ and $\lim_{n\to-\infty} h_n = 0$; if $\mu < 0$, then $\lim_{n\to\infty} h_n = 0$, and $\lim_{n\to-\infty} h_n = |\mu|^{-1}$.

The proof parallels that of Theorem 6.1. We first demonstrate the theorem when the generating function $\epsilon(u) = \sum_{-\infty}^{\infty} e_n u^n$ is rational. Consider the function

$$h_n(w) \;=\; \sum_1^{\infty} w^k e_n^{(k)} \tag{7.6}$$

for $0 < |w| < 1$. This series is absolutely convergent, since $e_n^{(k)}$ is a probability and $e_n^{(k)} \leqslant 1$. The generating function of $h_n(w)$ is $h(u,w) = w\,\epsilon(u)/\{1 - w\epsilon(u)\}$ and is a rational function. From (II.6.13) we have

$$h_n(w) \;=\; \frac{1}{2\pi i} \oint_{\substack{\text{unit}\\ \text{circle}}} w\,\epsilon(u)\{1 - w\,\epsilon(u)\}^{-1} u^{-(n+1)}\, du\,. \tag{7.7}$$

Suppose that $\mu = \epsilon'(1) > 0$. Then the circle of integration may be reduced slightly in radius to $|u| = R < 1$. We then infer that $h_n(w)$ is regular in w at $w = 1$, and that $h_n(1) = h_n = \sum_{k=1}^{\infty} e_n$. For h_n we have the representation

$$h_n \;=\; \frac{1}{2\pi i} \oint_{|u| = R} \frac{\epsilon(u)}{1 - \epsilon(u)} \cdot \frac{du}{u^{n+1}}\,. \tag{7.8}$$

When n is a large negative integer, the integrand of (7.8) may be evaluated by residues. All contributing singularities of $\epsilon(u)/\{1-\epsilon(u)\}$ have radii less than unity and it is clear that $\lim_{n\to-\infty} h_n = 0$. When n is a large positive integer, we first employ the transformation $\omega = u^{-1}$. The singularities in the ω plane which contribute all have radii less than 1† except that at $\omega = 1$ which contributes μ^{-1} to the integral. For rational $\epsilon(u)$ with $\mu < 0$, we use a circular contour of radius R slightly larger than 1 to verify the theorem. The divergence for $\mu = 0$ is verified as for Theorem 6.1.

Now consider the general $\epsilon(u)$ with $\mu > 0$. The reasoning is almost identical to that of Theorem 6.1. A direct corollary to the Lemma in Note 8 states that

$$|\{\epsilon(u) - 1 - \epsilon'(1)(u-1)\}/(1-u)^2| \tag{7.9}$$

is integrable over the unit circle when $\sum_{-\infty}^{\infty} |n| \log(1 + |n|)e_n < \infty$. One then confirms Theorem V.7.1 with the aid of the Riemann-Lebesgue Lemma. Details are given in Note 9.

Condition (A) of the theorem may be weakened to: (A′) If $\Sigma|n|e_n < \infty$ and $\mu \neq 0$. See Spitzer (1964), p. 282.

The extended renewal numbers h_n may be evaluated explicitly when $\epsilon(u)$ is sufficiently simple. The most important example is that for the lattice random variaole whose negative and positive increments both have geometrical distribution. Thus let

$$e_n = \begin{cases} q(1-\beta)\beta^{-n-1} & n < 0 \quad 0 < \beta < 1 \\ p(1-\alpha)\alpha^{n-1} & n > 0 \quad 0 < \alpha < 1 \\ 0 & n = 0 \quad p + q = 1, \end{cases} \tag{7.10}$$

so that

---

† *Lemma V.7.2.* If the magnitudes of possible increments, i.e., the set $\{|n| |\epsilon_n > 0\}$, have unity as their greatest common divisor, then there is exactly one zero of $\{1-\epsilon(u)\}$ on the unit circle located at $u = 1$.

For $\epsilon(u) = \Sigma\, e_n u^n$ with $\Sigma e_n = 1$, and $\epsilon(u) = 1$ for $|u| = 1$ implies that $u^n = 1$ for every possible increment.

$$\epsilon(u) = \frac{p(1-\alpha)u}{1-\alpha u} + \frac{q(1-\beta)}{u-\beta}, \tag{7.11}$$

and consider the case

$$\mu = \epsilon'(1) = p(1-\alpha)^{-1} - q(1-\beta)^{-1} > 0. \tag{7.12}$$

The renewal numbers may be obtained from (7.8). Since $\epsilon'(1) > 0$, the contour R must be chosen to be slightly smaller than 1. The function $1 - \epsilon(u)$ vanishes at $u = 1$ and at $\theta = (p\beta + q)/(p + q\alpha)$ for which $0 < \theta < 1$. When $n < 0$, we obtain at once

$$h_n = -\{\epsilon'(\theta)\}^{-1}\theta^{-(n+1)} \qquad n < 0. \tag{7.13a}$$

When $n > 0$, we observe from the transformation $u = w^{-1}$ that $\int_C \epsilon(u)\{1-\epsilon(u)\}^{-1} u^{-n-1}\,du$ is zero for any contour C containing $u = 0$, $u = \theta$, and $u = 1$ in its interior. Hence the integral of (7.8) is the negative of the integral for the same integrand and a contour of small radius about $u = 1$ in the positive sense. An additional term is present for $n = 0$. We find

$$h_n = \mu^{-1} \qquad n > 0 \tag{7.13a}$$

$$h_n = \mu^{-1} - \frac{p(1-\alpha)}{q\alpha + p} \qquad n = 0. \tag{7.13c}$$

V.8 A structural property of the extended renewal density and its analogue for the lattice

In Section 6 we saw that when $A(x)$ satisfies Theorem V.6.1 and has a non-vanishing first moment the extended renewal density $h(x)$ is asymptotically $|\mu|^{-1}$ in the direction of flow as $|x| \to \infty$. For Example A of that section, the exact equality $h(x) = \mu^{-1}$ held for all $x > 0$. We next show that this exact equality will be valid, when the negative increments have an exponential distribution, for any distribution of the positive increments.

Theorem V.8.1. If, for the increment distribution $A(x)$ with negative mean increment $\mu = \int x\,dA(x) < 0$,

$$A(x) = q\,e^{\eta x} \quad \text{for } x < 0, \tag{8.0}$$

then the extended renewal function $H(x) = |\mu|^{-1} x$ for all $x < 0$.

The proof is based on reasoning similar to that for Theorem 6.1. We first prove the theorem when: (a) positive increments are bounded; (b) $A(x)$ is absolutely continuous; (c) the density $a(x)$ has finite total variation on $(-\infty,\infty)$. From (a) and (8.0), the characteristic function $\alpha(z)$ is analytic at $z = 0$. Under the conditions stated, $\alpha(iV)$ will be monotonically increasing in V at $V = 0$ (Fig. V.1). The contour for

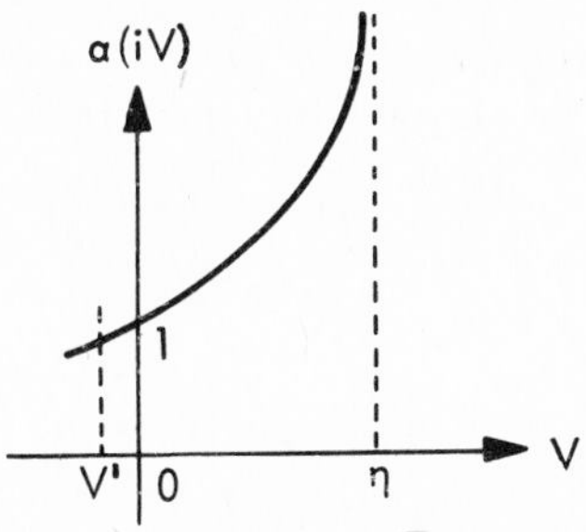

Fig. V.1 – Negative increments exponentially distributed. $\mu < 0$

the evaluation of $h(w,x)$ used in (6.3) may therefore be translated slightly to some $V' < 0$ at which $\alpha(iV') < 1$, and w may then be replaced by unity. To evaluate the extended renewal density,

$$h(x) = a(x) + \frac{1}{2\pi}\int_{-\infty+iV'}^{+\infty+iV'} \frac{\alpha^2(z)}{1-\alpha(z)}\, e^{-izx}\, dz$$

$$= \lim_{T\to\infty} \frac{1}{2\pi}\int_{-T+iV'}^{T+iV'} \frac{\alpha(z)}{1-\alpha(z)}\, e^{-izx}\, dz \qquad (8.1)$$

for $x < 0$ we close in the upper half-plane. Because of (8.0), condition (c), and Lemma III.3.1′, $|\alpha(z)|$ will fall off as $|z|^{-1}$ in the upper half-plane uniformly in $\arg z$, and Jordan's lemma is applicable. We may now invoke Rouché's theorem or the principle of the argument for the function $\{1-\alpha(z)\}$ to infer that there are as many poles as zeros inside the contour. Since $\{1-\alpha(z)\}$ has only one simple pole inside the contour at $z = i\eta$, it can have only one simple zero there and this is clearly the one at $z = 0$. By the method of residues, $h(x) = |\mu|^{-1}$ and the theorem follows for the restricted case. A more general distribution $A(x)$ will be the limit of some sequence of absolutely continuous distributions $A_j(x)$, all having the same mean, and all satisfying the conditions of the theorem and conditions (a), (b) and (c). Since $h_j(x) = |u|^{-1}$ for all $x < 0$ and all j, and since $H(x)$ for the limiting distribution exists, the theorem is proved.

When negative increments are exponentially distributed, but $\mu = \int x\, dA(x)$ is positive, the behaviour is slightly different.

Theorem V.8.2. If the increment distribution $A(x)$ of Theorem 8.1 with exponentially distributed negative increments has a positive mean increment $\mu = \int x\, dA(x) > 0$, then $H(x)$ is absolutely continuous for all $x < 0$ and its density there $h(x) = |\mu_{V_1}|^{-1} \exp(V_1 x)$. The value V_1 is that positive value in the convergence interval $(V_{\min}, V_{\max})$ at which $\alpha(iV_1) = 1$, and μ_{V_1} is the first moment of the conjugate distribution $A_{V_1}(x)$.

The theorem may be demonstrated with the aid of conjugate transformation. As for (6.11) it is clear that, for $|w| < 1$,

$$H(w,x) = \int_0^x e^{Vx'} d\left[\sum_1^{\infty} w^n \alpha^n(iV)\{A_V^{(n)}(x') - A_V^{(n)}(0)\}\right]. \qquad (8.1')$$

Under the assumptions of the theorem, the characteristic function has the form shown in Fig. V.2, and the value V_1 will always be in the

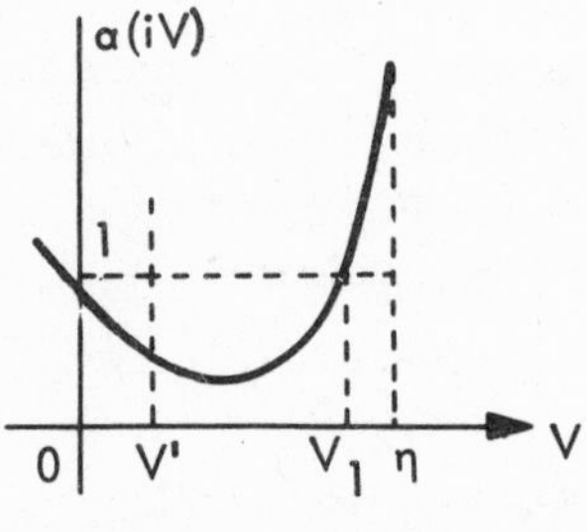

Fig. V.2—Negative increments exponentially distributed. $\mu > 0$

convergence interval. If we set $V = V_1$ in (8.1′), and $w = 1$ (the power series has a radius of convergence > 1 by the reasoning of Theorem 6.3), we have $H(x) = \int_0^x \exp(V_1 x')\, dH_{V_1}(x')$. But $A_{V_1}(x)$ satisfies the conditions of Theorem 8.1 and the result follows. The moment μ_{V_1} may be obtained from $\alpha(iV)$ by differentiation (III.2.2).

Now consider the extended renewal numbers h_n for the lattice. Lattice analogues of Theorem 8.1 and Theorem 8.2 follow.

Theorem V.8.3. If for the set of increment probabilities e_n with negative mean increment $\mu = \Sigma n e_n < 0$,

$$e_n = q(1-\beta)\beta^{-n-1} \quad \text{for } n < 0,$$

then $h_n = |\mu|^{-1}$ for all $n < 0$.

One first considers the case where positive increments are bounded so that $\epsilon(u)$ is analytic at $u = 1$. The proof is based on the notation and reasoning of Theorem 7.1. We may write

$$h_n = \frac{1}{2\pi i} \oint_{|u|=R} [1-\epsilon(u)]^{-1} u^{-(n+1)} du, \tag{8.2}$$

where R is slightly greater than 1, so that $\epsilon(R) < 1$. The g.f. $\epsilon(u) = \sum_0^\infty e_n u^n + q(1-\beta)(u-\beta)^{-1}$ has one singularity inside the circle $|u| = R$, and hence $1 - \epsilon(u)$ must have exactly one zero there by the principle of the argument. This zero is at $u = 1$, with residue $|\mu|^{-1}$ for all $n < 0$. For more general positive increments, a limiting procedure similar to that for Theorem 8.1 may be invoked.

It may be noted that the case $\beta = 0$ corresponds to the case where only unit increments occur in the negative direction. For this case we have the following small but significant modification.

Theorem V.8.3′. Under the assumption of Theorem 8.3 with $\beta = 0$, $h_n = |\mu|^{-1}$ for all $n < 0$, and $h_0 = |\mu|^{-1} - 1$. Consequently $g_n = \delta_{n,0} + h_n = |\mu|^{-1}$ for all $n \leqslant 0$. The proof is obvious. The same increment configuration with $\mu > 0$ is described by

Theorem V.8.4. If (a) $e_n = q(1-\beta)\beta^{-n-1}$ for $n < 0$; (b) $q > 0$, $\mu > 0$; then $h_n = |\mu_{\rho_1}|^{-1} \rho_1^{-n}$ for $n < 0$, where $\epsilon(\rho_1) = 1$, $0 < \rho_1 < 1$, and $\mu_{\rho_1} = \rho_1 \epsilon'(\rho_1)$.

For equation (8.2) is again valid with R slightly less than 1. There will be one zero of $1 - \epsilon(u)$ at $u = \rho_1$ on the positive real axis inside the circle, and the theorem follows from its residue and (III.6.5). The number μ_{ρ_1} is the mean of the conjugate distribution associated with ρ_1.

Theorem V.8.4′. Under the assumptions of Theorem 8.4 with $\beta = 0$, i.e., for a skip-free negative lattice process with mean positive increment, $g_n = \delta_{n,0} + h_n = |\mu_{\rho_1}|^{-1} \rho_1^{-n}$ for all $n \leqslant 0$.

The proof is again as above. Theorem V.8.3′ may also be proved as a direct consequence of the skip-free property, and Theorem V.8.4′ may be obtained via conjugate transformation (cf. Theorems V.8.2 and V.9.3).

V.9 The ergodic Green measure for the homogeneous drift processes

In this section we wish to examine the convergence of the ergodic Green measure

$$\mathbf{G}_c(x) = \int_0^\infty \{G(x,t) - G(0,t)\}\,dt \tag{9.0}$$

of (5.14) for the drift process $X(t)$ of Section II.3, and list some of its properties. In (9.0) $G(x,t)$ is the Green distribution (II.3.12) for the process $X(t)$ governed by increment distribution $A(x)$, frequency ν, and drift velocity v. The following theorem will be basic to the discussion.

Theorem V.9.1. If the increment distribution $A(x)$ governing the homogeneous drift process $X(t) = X_0 + \sum_1^{K(t)} \xi_i + vt$ of Section II.3 has a finite first moment μ, and if

$$\mu_V = \mu + \nu^{-1} v \neq 0, \tag{9.1}$$

the ergodic Green measure $\mathbf{G}_c(x)$ defined by (9.0) exists, is absolutely continuous in every finite interval, and its density $\mathbf{g}_c(x)$ is continuous for all $x \neq 0$. The ergodic Green density $\mathbf{g}_c(x) = (d/dx)\mathbf{G}_c(x)$ has the asymptotic behaviour:

$$\lim_{x\to\infty} \mathbf{g}_c(x) = (\nu\mu_V)^{-1}, \quad \lim_{x\to-\infty} \mathbf{g}_c(x) = 0 \quad \text{if } \mu_V > 0;$$

$$\lim_{x\to\infty} \mathbf{g}_c(x) = 0, \quad \lim_{x\to-\infty} \mathbf{g}_c(x) = |\nu\mu_V|^{-1} \quad \text{if } \mu_V < 0.$$

For consider the function

$$\tilde{G}(x,\theta) = \int_0^\infty G(x,t)\,e^{-\theta t}\,dt \tag{9.2}$$

for $\theta > 0$. The convergence of (9.2) is clear since $G(x,t) \leqslant 1$. Let $\tilde{\gamma}(z,\theta)$ be the Fourier-Stieltjes transform of $\tilde{G}(x,\theta)$. Absolute convergence permits the spatial integration to be carried out first, and we have

$$\tilde{\gamma}(z,\theta) = \int_0^\infty e^{-\theta t}\gamma(z,t)\,dt$$

$$= \int_0^\infty e^{-\theta t}\, e^{izvt} \sum_0^\infty \{e^{-\nu t}(\nu t)^k/k!\}\alpha^k(z)\,dt \qquad (9.3)$$

by (II.6.9). The integral-sum (9.3) is absolutely convergent for real z, and the order of summation and integration may be interchanged. Integrating with respect to t, we find

$$\tilde{\gamma}(z,\theta) = \nu^{-1}\sum_0^\infty \left(\frac{\nu}{\nu+\theta-iz\nu}\right)^{k+1}\alpha^k(z). \qquad (9.4)$$

$$= (\nu+\theta)^{-1}\left(\frac{\nu+\theta}{\nu+\theta-iz\nu}\right)\sum_0^\infty \left(\frac{\nu}{\nu+\theta}\right)^k\left\{\frac{\nu+\theta}{\nu+\theta-iz\nu}\,\alpha(z)\right\}^k.$$

From (9.4) it is easy to see that $\tilde{G}(x,\theta)$ is absolutely continuous. A comparison of (9.4) with (6.2) suggests that we may write

$$\tilde{g}(x,\theta) = (\nu+\theta)^{-1}\left\{\int_{-\infty}^\infty e_{\theta\nu}(x-x')\,dH\left(\frac{\nu}{\nu+\theta},x'\right) + e_{\theta\nu}(x)\right\}, \qquad (9.5)$$

where $E_{\theta\nu}(x)$ is the exponential distribution with $\mathfrak{F}-\mathfrak{S}$ transform $(\nu+\theta)(\nu+\theta-iz\nu)^{-1}$, and $H(w,x) = \sum_1^\infty w^k\{E_{\theta\nu}(x)\cdot A(x)\}^{(k)}$. The validity of (9.5) is assured, since $H(w,x)$ is of finite total variation on $(-\infty,\infty)$ when $|w| < 1$, and the integral in (9.5) converges. When $\theta \to 0$, $[H(\nu(\nu+\theta)^{-1},\xi)]_0^x \to H(x)$,† the extended renewal function (5.7) for the distribution $E_{0\nu}(x)\cdot A(x)$, provided that the mean of this distribution does not vanish. The latter condition is (9.1). We then have† from (9.5) that $\tilde{g}(x,\theta) \to \tilde{g}(x,0) = g_c(x)$ as $\theta \to 0$, where

$$g_c(x) = \nu^{-1}\left\{\int_{-\infty}^\infty e_{0\nu}(x-x')\,dH(x') + e_{0\nu}(x)\right\}. \qquad (9.6)$$

The distribution $E_{0\nu}(x)\cdot A(x)$ is absolutely continuous and of finite total variation.‡ Hence the function $H(x)$ will be absolutely continu-

† The limiting operations are justified by monotonicity.

‡ For $A(x)$ may be represented as the limit of a sequence of absolutely continuous distributions $A_j(x)$, and Lemma II.3.1 may be employed.

ous on every finite interval, and by Theorem 6.1† one will have $h(x) \sim |\mu_\nu|^{-1}$ in one direction and $h(x) \to 0$ in the other. The remaining behaviour of $g_c(x)$ stated in the theorem follows from the structure of (9.6).

We note that when $v \to 0$, $E_{0v}(x) \to U(x)$ and (9.5) is consistent with (5.9) and (5.14).

The kind of structural property of the extended renewal density described in Theorems 8.1 and 8.2 is also available for the ergodic Green density $g_c(x)$ when the distribution $A(x)$ has positive support only, and the drift rate v is negative. We have specifically the following theorems.

Theorem V.9.2. Let $A(x)$ be an increment distribution of positive support. If the drift velocity $v < 0$ and $\mu_v = \mu + \nu^{-1} v < 0$, the ergodic Green density $g_c(x)$ defined by Theorem (9.1) has the value $\nu^{-1}|\mu_v|^{-1}$ for all $x < 0$.

The proof of the theorem runs parallel to that for Theorem 8.1. One first considers distributions $A(x)$ for positive bounded increments. From (9.4) we may write

$$\tilde{g}(x,\theta) = \frac{1}{2\pi} \lim_{T\to\infty} \int_{-T}^{T} e^{-izx} \left[\frac{1}{\nu+\theta-izv} \left\{ \frac{1}{1 - \frac{\nu}{\nu+\theta}\,\frac{\nu+\theta}{\nu+\theta-izv}\,\alpha(z)} \right\} \right] dz. \tag{9.7}$$

By our assumptions the denominator of the expression in curly brackets is analytic at $z = 0$. Let θ be such that $\mu + (\nu+\theta)^{-1} v < 0$. For this value θ, we may translate the contour of integration of (9.7) slightly to some new contour $(-\infty-i\tau, \infty-i\tau)$ with $\tau > 0$, without changing the value of the integral (9.7). Now let τ be fixed and let θ pass to zero through the intervening values $0 < \theta' < \theta$. We then have

$$g_c(x) = \frac{1}{2\pi} \lim_{T\to\infty} \int_{-T-i\tau}^{+T-i\tau} e^{-izx} \left[\frac{1}{\nu-izv} \left\{ \frac{1}{1 - \frac{\nu}{\nu-izv}\,\alpha(z)} \right\} \right] dz. \tag{9.8}$$

For $x < 0$ we may now close in the upper half-plane. The function $\alpha(z)$ is analytic in this half-plane and the denominator of the expression in curly brackets has a simple pole at $z = -i\nu/v$. Hence by Rouché's theorem there is one zero for the denominator, and this is the one known to be at $z = 0$. The theorem follows in the restricted case

† When $A(x)$ has only a first moment, see Note 8.

from the residue at zero. For arbitrary $A(x)$ with positive support, a sequence of distributions $A_j(x)$ with bounded increments and $\mu_j = \mu$ may be employed to complete the proof as for Theorem 8.1.

We will now demonstrate the analogue of Theorem 8.2 for the homogeneous drift processes.

Theorem V.9.3. Let $A(x)$ have positive support only, drift rate $v < 0$, and $\mu_V = \mu + \nu^{-1} v > 0$. Then the ergodic Green density $g_C(x)$ defined by Theorem 9.1 is given for all $x < 0$ by $g_C(x) = |\mu_{V_1}(1)|^{-1} \exp(V_1 x)$. The value V_1 is that positive value always available in the convergence interval of $G(x,1)$ at which $\gamma(iV,1) = 1$; $\mu_{V_1}(1)$ is the first moment of the conjugate distribution $G_{V_1}(x,1)$, and is given from (III.7.5) by $\mu_{V_1}(1) = \nu\alpha(iV_1)\mu_{V_1} + v$.

The theorem is obtained from Theorem 9.2 and conjugate transformation, and the reasoning required is a direct analogue of that for Theorem 8.2. Under the conditions of the theorem, it is clear that a family of distributions $G_V(x,t)$ conjugate to $G(x,t)$ will be available for every value V in the interval $[0,\infty)$. A plot of $\gamma(iV,t) = \exp\{-\nu t + \nu t\alpha(iV) - vVt\}$ against V for $t = 1$ might have the form shown in Fig. V.3.

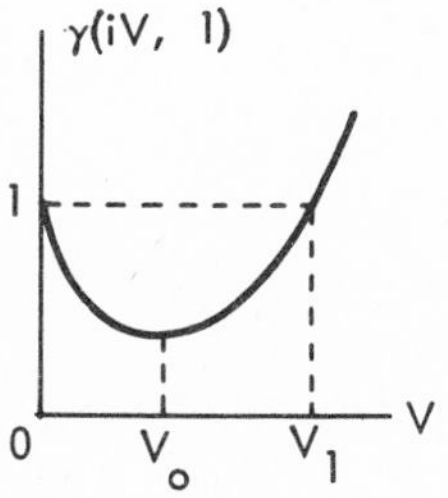

Fig. V.3

The function

$$\tilde{G}_V(x,s) = \int_0^\infty e^{-st}\{G_V(x,t) - G_V(0,t)\}\,dt$$

defined in (9.2) may be rewritten with the aid of the conjugate transformation III.2.1 in the form

$$\tilde{G}_{V_A}(x,s) = \int_0^x \exp\{(V_B - V_A)x'\}\, dG\left(s + \left[\nu\alpha(iV) - vV\right]_{V_A}^{V_B}, x'\right). \tag{9.9}$$

Since $\nu\alpha(iV) - vV > \nu\alpha(iV_0) - vV_0$ when $V \neq V_0$, it is readily seen that the Laplace transform $\tilde{G}_V(x,s)$ has a negative abscissa of convergence for every $V \neq V_0$. In particular, we have from (9.9) for $V_A = 0$, and $V_B = V_1$ when $s = 0$, $\mathbf{G}_C(x) = \int_0^x e^{V_1 x'} d\mathbf{G}_{V_1 C}(x')$. Theorem V.9.3 then follows from Theorem V.9.2.

V.9.1 The ergodic Green measure for more general processes in continuous time

Consider the homogeneous process of (II.3.20),

$$X(t) = X_J(t) + X_D(t) + vt \tag{9.10}$$

where: (a) $X_J(t)$ is a jump process with increments of distribution $A(x)$ and frequency ν commencing at $X_J(0) = 0$; (b) $X_D(t)$ is a Wiener-Lévy process (II.3) with parameter $D > 0$. The Green distribution $G(x,t)$ is the convolution of that for $X_J(t) + vt$ and $X_D(t)$, and will inherit the absolute continuity of $G_D(x,t)$ for $t > 0$. The convergence and simple asymptotic properties of the ergodic Green density $\mathbf{g}_C(x) = \int_0^\infty g(x,t)\,dt$ for $X(t)$ may now be stated.

Theorem V.9.4. The integral $\mathbf{g}_C(x) = \int_0^\infty g(x,t)\,dt$ defining the ergodic Green density for the process (9.10) converges for every value of x when at $t = 1$, $X(1)$ has a finite, non-vanishing mean

$$\mu(1) = \int x\,g(x,1)\,dx \neq 0. \tag{9.11}$$

The ergodic Green density $\mathbf{g}_C(x)$ is continuous and uniformly bounded on $(-\infty,\infty)$. If $\mu(1) > 0$, $\lim_{x\to\infty} \mathbf{g}_C(x) = \{\mu(1)\}^{-1}$ and $\lim_{x\to-\infty} \mathbf{g}_C(x) = 0$.

As proof, we observe that since $g(x,t)$ is infinitely divisible (II.3d), we may write formally

$$\mathbf{g}_C(x) = \int_0^\infty g(x,t)\,dt = \left\{\int_0^1 g(x,t)\,dt\right\}\cdot\left\{\delta(x) + \sum_1^\infty g^{(k)}(x,1)\right\}, \tag{9.12}$$

where the dot denotes convolution of the expressions in brackets and $g^{(k)}(x,1)$ is the k-fold convolution of the probability density $g(x,1)$. The function $\gamma(x) = \int_0^1 g(x,t)\,dt$ is also a probability density. Moreover, $g(x,1)$ is continuous and of finite total variation by virtue of these properties for $g_D(x,1)$. From Theorem 6.1, the series $h(x,1) = \sum_1^\infty g^{(k)}(x,1)$ converges, is continuous, $\lim_{x\to\infty} h(x,1) = \{\mu(1)\}^{-1}$, and $\lim_{x\to-\infty} h(x,1) = 0$. The convolution between the two bracketed expressions in (9.12) is therefore seen to converge, and the representation (9.12) is justified by this convergence and the positive character of all elements contributing to the summation and convolution (Fubini's theorem). The theorem then follows.

The structural property for the drift processes exhibited in Theorem 9.2 is unaffected by the addition of a Wiener-Lévy component. The following theorem is easily demonstrated.

Theorem V.9.5. If, for the process (9.10), $X_J(t)$ has positive increments, and $\mu(1) = \int x\, g(x,1)\, dx < 0$, then the ergodic Green density $\mathbf{g}_C(x)$ has the value $|\mu(1)|^{-1}$ for every $x \leqslant 0$.

This property is a direct consequence of the skip-free character of the process. If one refers to (IV.5.2) and integrates with respect to t from 0 to ∞, one obtains $\mathbf{g}_C(-x-y) = \mathbf{g}_C(-x) S_y(\infty)$ for every $x > 0$ and $y > 0$. But $\mu(1) < 0$ implies that† $S_y(\infty) = 1$. We then conclude that $\mathbf{g}_C(x)$ is constant for $x < 0$, and the theorem follows from Theorem 9.4.

V.10 Asymptotic representations of the extended renewal function

In Chapter III, a convenient set of saddlepoint approximations for the time-dependent Green's functions were exhibited when the characteristic function $\alpha(z)$ of the governing increment distribution had a suitable convergence strip. A similar procedure is available under comparable conditions on $A(x)$ for the extended renewal density $h(x)$ and hence for the ergodic Green density $\mathbf{g}(x)$.

For suppose the increment distribution $A(x)$ has a conjugate set of distributions $A_V(x)$ (Section III.2) containing the value V_0 at which $\alpha(iV)$ takes on its minimum. If $A(x)$ is absolutely continuous, we have from (III.2.1) $a(x) = \alpha(iV)\, e^{Vx} a_V(x)$ and $a^{(n)}(x) = \alpha^n(iV) \times e^{Vx} a_V^{(n)}(x)$. In particular for $V = V_0$, we have from $h(x) = \sum_1^\infty a^{(k)}(x)$,

$$h(x) = \exp(V_0 x) \sum_1^\infty \alpha^k(iV_0)\, a_{V_0}^{(k)}(x). \tag{10.1}$$

† This familiar result may be inferred simply from the Central Limit Theorem.

We note that $\alpha(iV_0) < 1$. Moreover $\mu_{V_0} = 0$ and $A_{V_0}(x)$ is the conjugate distribution corresponding to the zone of convergence fixed about the origin (Section III.5.1). Hence for x fixed we have the asymptotic representation as $k \to \infty$ $a_{V_0}^{(k)}(x) \sim \exp(-x^2/2k\sigma_{V_0}^2)/(2\pi k\sigma_{V_0}^2)^{\frac{1}{2}}$, and the convergence of the series (10.1) is assured. When $\alpha(iV_0) \ll 1$, the convergence of (10.1) for x near the origin will be rapid. The Central Limit approximation to $a_{V_0}^{(k)}(x)$ for large k may be advantageous.

We have seen in Section 6 that under certain simple conditions when $A(x)$ has a negative first moment $\mu < 0$, the asymptotic behaviour of $h(x)$ for large negative x is $h(x) \sim |\mu|^{-1}$ as $x \to \infty$. For large positive x, when $\mu < 0$, we saw that $\lim_{x\to\infty} h(x) = 0$. The asymptotic behaviour in the positive direction may be sharpened under the following conditions.

Theorem V.10.1. Let $a(x)$ be a probability density with a negative first moment for which (1) $a^{(m)}(x)$ is bounded for some m; (2) the convergence interval for $\alpha(z)$ has in its interior the point $z = iV_1$ with $V_1 < 0$ at which $\alpha(iV_1) = 1$; (3) $a(x)\exp\{-V_1 x\}$ goes to zero as $x \to +\infty$; then

$$h(x) \sim \mu_{V_1}^{-1} e^{V_1 x} \quad \text{as } x \to \infty. \tag{10.2}$$

The proof is as follows. As for (10.1) we have $h(x) = e^{V_1 x} \times \sum_1^\infty \alpha^n(iV_1) a_{V_1}^{(k)}(x) = e^{V_1 x} h_{V_1}(x)$. The boundedness of $a^{(m)}(x)$ and condition (2) imply that $a_{V_1}^{(m)}(x) \in L_{1+\delta}$ for some $\delta > 0$, and hence that $a_{V_1}^{(q)}(x)$ is bounded and continuous for some q. We may now apply Theorem V.6.2 to infer that $h_{V_1}(x) \sim \mu_{V_1}^{-1}$ as $x \to \infty$, since $\mu_{V_1} > 0$, and the theorem follows.†

If the positive increments are exponentially distributed, equality holds for (10.2) (Theorem V.8.2). An asymptotic series for $h(x)$ whose first term is (10.2) is often available. When $A(x)$ has positive support the series is known as the Lotka expansion (Feller, 1941). A theorem representative of this type of result obtained with the aid of conjugate transformation is the following.

---

† The parameters μ_{V_1} and V_1 are easy to exhibit analytically when the distribution $A(x)$ is conjugate to a distribution $A_0(x)$ symmetric about $x = 0$, with $A(x) = A_V(x)$ obtained from (III.2.1). Then it is easy to see that $V_1 = -2V$ and that $\mu_{V_1} = |\mu_V|$.

Theorem V.10.2. Let $a(x)$ be a probability density with a negative first moment for which (1) $a(x)$ is bounded; (2) the c.f. $\alpha(z)$ has a convergence strip with lower boundary $V_{\min} < 0$. Let the zeros of $\{1 - \alpha(z)\}$ in the strip $V_{\min} < V < 0$ be simple and designated by iV_1, $iV_2 \pm U_2$, etc., with $0 > V_1 > V_2 > \ldots$. Then if $V_{\min} < 2V_k$, $h(x) - a(x)$ has the asymptotic expansion to k terms

$$h(x) - a(x) \sim \frac{1}{\mu_{V_1}} e^{V_1 x} - i \sum_{j=2}^{k} \left\{ \rho_{j+} e^{V_j x - iU_j x} + \rho_{j-} e^{V_j x + iU_j x} \right\} \quad \text{as } x \to +\infty, \qquad (10.2')$$

where ρ_{j+} and ρ_{j-} are the residues of $\alpha^2(z)\{1 - \alpha(z)\}^{-1}$ at $iV_j + U_j$ and $iV_j - U_j$ respectively.

The proof is as follows. When $a(x)$ is bounded, $a(x)$ and $\alpha(U) \in L_2$. Moreover, $a(x)\,e^{-Vx}$ and hence $a_V(x) \in L_2$ for all $0 \geqslant V > \frac{1}{2}V_{\min}$. Consequently, for fixed V in $[0, \frac{1}{2}V_{\min})$, $\alpha(U + iV) = \alpha(iV)\alpha_V(U) \in L_2$ and by the Riemann-Lebesgue lemma, $\alpha(\pm\infty + iV) = 0$. For $h(w,x) = \sum_1^\infty w^n a^{(n)}(x)$, we have $h(w,x) - w\,a(x) = (2\pi)^{-1} \int_{-\infty}^{\infty} w^2 \alpha^2(z)\{1 - w\,\alpha(z)\}^{-1} \times e^{-izx} dz$ when $0 \leqslant |w| < 1$. Hence, as for (8.1),

$$h(x) - a(x) = \frac{1}{2\pi} \int_{iV-\infty}^{iV+\infty} \frac{\alpha^2(z)}{1 - \alpha(z)} e^{-izx}\, dz, \qquad V_1 < V < 0.$$

From Cauchy's theorem, we then have

$$h(x) - a(x) = \frac{1}{\mu_{V_1}} e^{V_1 x} + \sum_{j=2}^{k} \left\{ \rho_{j+} e^{(V_j + iU_j)x} + \rho_{j-} e^{(V_j - iU_j)x} \right\}$$

$$+ \frac{1}{2\pi} \int_{iV-\infty}^{iV+\infty} \frac{\alpha^2(z)}{1 - \alpha(z)} e^{-izx}\, dz; \qquad V_{k+1} < V < V_k;\ \tfrac{1}{2}V_{\min} < V.$$

But $\alpha(U+iV) = \alpha(iV)\,\alpha_V(U)$. Hence the magnitude of the remainder term is less than $\left\{ \frac{1}{2\pi} \int_{-\infty}^{\infty} \left| \frac{\alpha^2(iV)\alpha_V^2(U)}{1 - \alpha(iV)\alpha_V(U)} \right| dU \right\} e^{Vx}$ and this is a higher order infinitesimal, as $x \to \infty$, than the kth term of the asymptotic expansion. Q.E.D.

V.10.1 Asymptotic decay of other ergodic Green's functions

The asymptotic behaviour represented in Theorem V.10.1 is also present for the homogeneous process (9.10) in continuous time. For let

$$X(t) = X_J(t) + X_D(t) + vt\,, \tag{10.3}$$

where $X_J(t)$ is the jump process with Poisson increments of frequency ν and distribution $A(x)$ and $X_D(t)$ is the Wiener-Lévy process. For $Dt > 0$, $G(x,t)$ will be absolutely continuous and we may write as in Section III.7 $g_V(x,t) = g(x,t)e^{-Vx}/\gamma(iV,t)$ when a convergence strip is available. We then have

$$g(x,t) = e^{Vx}\exp\left[-\nu t\{1-\alpha(iV)\} - vVt + DV^2t\right] g_V(x,t), \tag{10.4}$$

where $g_V(x,t)$ is the Green density for the process $X_V(t)$ conjugate to (10.3). We saw in Theorem III.7.1 that when $A(x)$ has a complete characteristic function, the characteristic function of $G(x,t)$ will also be complete. Consequently we anticipate that when $G(x,1)$ has a negative first moment $\mu(1) = v + \nu\mu < 0$ the expression in square brackets in (10.4) will vanish for some $V = V_1 < 0$. By Theorem 9.4, $X(t)$ will have an ergodic Green density $g_C(x) = \int_0^\infty g(x,t)\,dt$, and since $\mu_{V_1}(1) = \int x\, g_{V_1}(x,1)\,dx > 0, X_{V_1}(t)$ will have an ergodic Green density $g_{CV_1}(x) = \int_0^\infty g_{V_1}(x,t)\,dt$. Hence from (10.4),

$$g_C(x) = \exp(V_1 x)\, g_{CV_1}(x)\,. \tag{10.5}$$

It may then be inferred from Theorem 9.4 that in the direction of decay, $g_C(x) \sim \{\mu_{V_1}(1)\}^{-1}\exp(V_1 x)$. If $D = 0$ but $v \neq 0$, an ergodic density is still available and (10.5) is still valid. We note from (III.7.4), and (III.7.5) that

$$\mu_{V_1}(1) = \nu\alpha(iV_1)\mu_{V_1} + v - 2DV_1\,. \tag{10.5$'$}$$

When $D = 0$, this reduces to $\mu_{V_1}(1) = v + \nu\alpha(iV_1)\mu_{V_1}$. In summary we have the following theorem.

Theorem V.10.3. If for the homogeneous process $X(t)$ of (10.3): (a) $G(x,1)$ has a negative first moment; (b) the increment distribution $A(x)$ governing the jump process $X(t)$ has a complete characteristic function; (c) either $D > 0$, or $v \neq 0$, or both; then for the ergodic Green density $g_C(x)$ we have, in the direction of decay,

$$g_c(x) \sim \{\mu_{V_1}\}^{-1} \exp(V_1 x), \quad \text{as } x \to \infty, \tag{10.6}$$

where iV_1 with $V_1 < 0$ is the point on the imaginary axis in the convergence strip at which

$$-\nu\{1 - \alpha(iV_1)\} - vV_1 + DV_1^2 = 0 \tag{10.7}$$

and $\mu_{V_1}(1)$ is given by (10.5′).

The reader may verify that the relation (10.6) is not merely asymptotic but exact when $A(x)$ has no positive support.

Direct analogues of equations (10.1) and (10.2) may be exhibited for the lattice. Let the increment probabilities e_n with generating function $\epsilon(u)$ have $\mu = \Sigma\, n\, e_n \neq 0$. If $\epsilon(u)$ has a convergence annulus containing the value ρ_0 at which $\epsilon(\rho)$ has a minimum, then for $h_n = \sum_1^\infty e_n^{(k)}$, we have from (III.6.3) $e_n^{(k)} = \rho_0^{-n}\, \epsilon^k(\rho_0)\, e_{\rho_0;n}^{(k)}$, and (10.1) becomes

$$h_n = \rho_0^{-n} \sum_1^\infty \epsilon^k(\rho_0)\, e_{\rho_0;n}^{(k)}. \tag{10.8}$$

The direct analogue of Theorem V.10.1 follows.

Theorem V.10.4. If the increment probabilities e_n have a negative mean $\mu = \Sigma\, n\, e_n$, and the generating function $\epsilon(u)$ has an annulus of convergence containing the value $\rho_1 > 1$ at which $\epsilon(\rho_1) = 1$, then

$$h_n \sim \frac{1}{\mu_{\rho_1}} \rho_1^{-n} \quad \text{as } n \to \infty, \tag{10.9}$$

provided that for $n \geqslant N_0$, all states n are accessible in the sense of Theorem 7.1.

The proof is straightforward. The only point worth noting is that if a state n is accessible for a lattice process N_k, it will be accessible for every conjugate process $N_{\rho k}$ associated with the convergence strip, and in particular for the process with $\rho = \rho_1$.

V.11 Approach to ergodicity

Consider an ergodic process X_k with distributions $F_k(x)$ and compensation functions $C_k(x)$ related by (2.1) and (2.2). We have seen that the functions $C_k(x)$ and $C_\infty(x)$ may be represented as the difference between two distribution functions. Consequently, these compensation functions will have Fourier-Stieltjes transforms

$$\psi_k(z) = \int e^{izx}\, dC_k(x) \tag{11.1}$$

for z real. The transform of (2.3) is

$$\phi_k(z) = \sum_{k'=0}^{k} \alpha^{k-k'}(z)\, \psi_{k'}(z). \tag{11.2}$$

The generating functions $\phi(w,z) = \sum_0^\infty w^k \phi_k(z)$ and $\psi(w,z) = \sum_0^\infty w^k \psi_k(z)$ are well defined for z real and $|w| < 1$. We then have from (11.2) after multiplication by w^k and summation,

$$\phi(w,z) = \frac{\psi(w,z)}{1 - w\alpha(z)}. \tag{11.3}$$

If we multiply (11.3) by $(1-w)$ and let $w \to 1$ along any path in the interior of the unit circle, we have, for $\alpha(z) \neq 1$,

$$\lim (1-w)\, \phi(w,z) = \{1-\alpha(z)\}^{-1} \lim (1-w)\, \psi(w,z). \tag{11.4}$$

Because of the ergodicity we know that $\phi_k(z) \to \phi_\infty(z)$ and $\psi_k(z) \to \psi_\infty(z)$ as $k \to \infty$. Consequently the two limits† are $\phi_\infty(z)$ and $\psi_\infty(z)$, and we have for the characteristic function of the ergodic distribution

$$\phi_\infty(z) = \frac{\psi_\infty(z)}{1-\alpha(z)}. \tag{11.5}$$

Consider next the function $\mathbf{G}(w,x) = \sum w^k\{A^{(k)}(x) - A^{(k)}(0)\}$ for $0 < |w| < 1$. The function $\mathbf{G}(w,x)$ is of bounded variation and has a Fourier-Stieltjes transform $\{1 - w\alpha(z)\}^{-1}$ for real z; the function $F(w,x) = C_\infty(x) \cdot \mathbf{G}(w,x)$ is also of bounded variation and has the Fourier-Stieltjes transform $\phi(w,z) = \psi_\infty(z)\{1 - w\alpha(z)\}^{-1}$. Since $\phi(w,z) \to \phi_\infty(z)$ as $w \to 1-$, it is clear that $F(w,x) \to F_\infty(x)$ and that $F_\infty(x) - C_\infty(x) \cdot \mathbf{G}(1,x)$ must be a constant. We then have (5.8), i.e.

$$F_\infty(x) - F_\infty(0) = [C_\infty(x) \cdot \mathbf{G}(x)]_0^x. \tag{11.6}$$

† This statement follows from Abel's theorem on the continuity of power series. For let $a_n \to a_\infty$ as $n \to \infty$. Then for $|w| < 1$, $\alpha(w) = \sum_0^\infty a_n w^n = a_\infty(1-w)^{-1} + \sum_0^\infty b_n w^n$ where $b_n = a_n - a_\infty \to 0$ as $n \to \infty$. For $|w| < 1$, $(1-w)\sum_0^\infty b_n w^n = b_0 + (b_1 - b_0)w + (b_2 - b_1)w^2 \ldots$. Since this series converges to zero at $w = 1$, $\lim(1-w)\beta(w) = 0$ as $w \to 1$ from inside the circle by Abel's theorem.

For a process in continuous time for which (3.3) is valid and for which $C(x,t)$ approaches a limiting value $C(x,\infty)$, the same type of discussion is available. The Laplace transform with respect to t is found from (3.3) to be

$$\phi^*(z,s) = \phi_0(z)\,\gamma^*(z,s) + \psi^*(z,s)\,\gamma^*(z,s), \tag{11.7}$$

where $\gamma^*(z,s)$ is the combined Laplace and Fourier-Stieltjes transform of the Green distribution $G(x,t)$.† For the homogeneous process (II.3.20), we have from (II.7.3),

$$\gamma(z,t) = \exp[-\nu t\{1-\alpha(z)\} - Dz^2 t + iz v t] \tag{11.8}$$

so that for real z and $\mathrm{Re}(s) > 0$,

$$\gamma^*(z,s) = [s + \nu\{1-\alpha(z)\} + Dz^2 - izv]^{-1}. \tag{11.9}$$

If we multiply (11.7) by s and let s approach zero through the half-plane $\mathrm{Re}(s) > 0$, we obtain, by the usual Tauberian argument, as a consequence of the ergodicity and limiting behaviour of $C(x,t)$ assumed,

$$\phi_\infty(z) = \psi_\infty(z)\,\gamma^*(z,0) = \frac{\psi_\infty(z)}{\nu\{1-\alpha(z)\} + Dz^2 - izv}. \tag{11.10}$$

From this equation, the identification (5.13) follows as above.

For the homogeneous lattice processes, consider equation (4.7). Let

$$\psi^{(k)}(u) = \sum_{-\infty}^{\infty} c_n^{(k)} u^n. \tag{11.11}$$

The convergence of this series on the unit circle $|u| = 1$ is assured by (4.8). The transform of (4.7), valid on the unit circle, is

$$\pi^{(k)}(u) = \pi^{(0)}(u)\,\epsilon^k(u) + \sum_{k'=1}^{k} \psi^{(k')}(u)\,\epsilon^{(k-k')}(u). \tag{11.12}$$

For the generating functions $\pi(w,u) = \sum_0^\infty w^k \pi^{(k)}(u)$, and $\psi(w,u) = \sum_0^\infty w^k \psi^{(k)}(u)$, defined for $|u| = 1$ and $|w| < 1$, we then find

† The reader is asked to bear with the apparent inconsistency between the asterisk representation of Laplace transforms here employed, and the tilde representation of Section 9, there resorted to with reluctance.

$$\pi(w,u) = \frac{\psi(w,u)}{1 - w\,\epsilon(u)}. \tag{11.13}$$

Ergodicity then implies, as above, that the ergodic probabilities $p_{\infty n}$ have the generating function

$$\pi_\infty(u) = \frac{\psi_\infty(u)}{1 - \epsilon(u)}. \tag{11.14}$$

The validity of (7.3) as a limiting form may be inferred from (11.14) by the previous argument.

V.12 Wiener-Hopf and Hilbert problems

Consider the replacement process X_k (Section V.1) on the semi-infinite interval $[0,\infty)$ governed by increment distribution $A(x)$ and replacement distribution $B(x)$. The generating function with respect to k and Fourier-Stieltjes transform with respect to x of $F_k(x)$ is given by (11.3). If all replacements are to the state $x = 0$, $B(x) = U(x)$ and the compensation functions $C_k(x)$ will have no positive support when $k \geqslant 1$. Consequently, $\psi_k(z)$ will be analytic in the lower half-plane $V = \text{Im}\,(z) < 0$ and for $|w| < 1$, $X(w,z) = \sum_1^\infty w^k \psi_k(z)$ will also be analytic there. Equation (11.3) then reads

$$\{1 - w\,\alpha(z)\}\, \phi^+(w,z) = \phi_0^+(z) + X^-(w,z). \tag{12.1}$$

Positive superscripts have been added to denote analyticity in the half-plane $V > 0$, and negative superscripts to denote analyticity in the half-plane $V < 0$. Equation (12.1) has the form of a Wiener-Hopf problem, or more properly a Hilbert problem on the real line $V = 0$.

A Wiener-Hopf discussion of the waiting-time process of the first example in Section 1 was given by Lindley (1951) and amplified subsequently by Smith (1953). A history of earlier discussions of the Wiener-Hopf problem for bounded homogeneous processes may be found in Kemperman (1961). A full discussion of such Hilbert problems is available in the treatise of Muskhelishvili (1953). A simplified discussion of the particular problem (12.1) is given elsewhere (Keilson, 1961). We will confine our discussion here to the character of (12.1) and some nomenclature.

The function $\{1 - w\,\alpha(z)\}$ is called the *kernel* of the Wiener-Hopf problem. The kernel may only be defined on the line $V = 0$, since the distribution function $A(x)$ is arbitrary and its convergence strip may have zero width. The remaining functions in (12.1) are defined in half-planes of convergence which may only have the line $V = 0$ in common. If $X_0 = 0$, $\phi_0^+(z) = 1$ and (12.1) takes the form

$$\{1 - w\alpha(z)\} = \Psi^-(w,z)/\Psi^+(w,z) \tag{12.2}$$

where (a) $\Psi^-(w,z) = 1 + X^-(w,z)$ is analytic in the half-plane $V < 0$ and $\Psi^+(w, U+i\infty)$ has a non-zero value independent of U, and (b) $\Psi^+(w,z) = \phi^+(w,z)$ is analytic in the half-plane $V > 0$, has a non-zero value at $V = +\infty$ independent of U, and $\Psi^+(w,0) = (1-w)^{-1}$. When $|w| < 1$, the real part of the kernel is positive on the line $V = 0$. The theory of the Hilbert problem then states that the factorization of the kernel exhibited in (12.2) subject to the stated conditions on Ψ^- and Ψ^+ is unique. The *homogeneous Hilbert problem* requests the display of such factors for a given kernel.

We next consider an ergodic process X_k, with ergodic distribution $F_\infty(x)$. For an arbitrary replacement distribution $B(x)$, it follows from (5.2) and (1.6b) that $C_\infty(x) = K[Z(x) + B(x)]$, where K is some undetermined constant, $Z(x)$ has non-positive support and $B(x)$ has non-negative support. We then have $\psi_\infty(z) = K\{\zeta^-(z) + \beta^+ z)\}$, and equation (11.5) states that

$$\{1 - \alpha(z)\}\, \phi_\infty^+(z) = K\{\zeta^-(z) + \beta^+(z)\}. \tag{12.3}$$

Here the kernel $\{1 - \alpha(z)\}$ vanishes at $z = 0$, and the ordinary Hilbert theory must be modified. Because of the presence of the term $\beta^+(z)$ on the right, (12.3) is said to be an *inhomogeneous Hilbert problem.*

When the process has its state space on the lattice of non-negative integers, a formulation in terms of the generating function is more convenient. The lattice analogue of (11.3) is (11.13). To obtain the lattice analogue of (12.1) we rewrite (11.13), setting $X(w,u) = \sum_1^\infty \psi_k(u) w^k$, as

$$\{1 - w\,\epsilon(u)\}\, \pi^+(w,u) = \pi^{+(0)}(u) + X^-(w,u). \tag{12.4}$$

Here a positive superscript denotes analyticity in u in the interior of the unit circle $|u| = 1$, and a negative superscript analyticity in the exterior of the unit circle. Equality in (12.4) need only hold on the unit circle, and the equation constitutes a Hilbert problem on that contour. A discussion of (12.4) may be found in Keilson (1961) where the process $N_k = \max[0, N_{k-1} + \xi_k]$ is discussed in a queuing theory context. A treatment of (12.4) may also be found in Keilson (1962c).

For processes on the finite interval, a modification of the Wiener-Hopf discussion has been given by Kemperman (1964).

The answers obtained by these methods are often formal and offer little insight into the statistical behaviour of the process. In simple instances the factorization (12.2) basic to the homogeneous and inhomogeneous problems may be exhibited in closed form. More generally, this factorization requires the location of zeros of the kernel in the complex plane. If the pertinent zeros are not few in number, or conveniently located, this task is fruitless.

The methods are important, however, as analytical tools. In Note 10 they are employed to derive an important identity due to Spitzer and the related results of E. Sparre Andersen.

CHAPTER VI

ERGODIC DISTRIBUTIONS, RUIN PROBLEMS, AND EXTREMES

The results of this chapter are employed to derive simple representations of the ergodic distributions for a variety of important homogeneous continuum and lattice processes modified by either a single retaining boundary or two retaining boundaries. Related replacement processes are treated in the same way. The simple asymptotic behaviour of the ergodic Green's function gives rise to a corresponding simplicity of behaviour of the ergodic distributions away from the boundaries. The gambler's ruin probabilities for the homogeneous processes on a finite interval may be obtained directly from the ergodic distributions. Consequently, the distribution of the greatest value attained between returns to zero for ergodic processes with a single retaining barrier may be exhibited at once.

VI.1 Conditions for ergodicity

For homogeneous processes modified by one or two retaining boundaries (Section V.1), the stationary distributions associated with ergodicity are readily exhibited when the structural property developed in Sections V.8 and V.9 is available. Before discussing the various cases of interest, a brief discussion of the conditions for ergodicity (V.5) is in order.

For any homogeneous process in discrete time modified by a single retaining boundary, ergodicity is assured when the homogeneous process has a net motion toward the boundary. This is stated in the following theorem due to Lindley (1952) which we give without proof.

Theorem VI.1.1. Consider the process X_k defined by $X_0 > 0$, and the recursion law $X_{k+1} = \max[0, X_k + \xi_{k+1}]$ where ξ_k is a sequence of independent random variables assuming both positive and negative values. A sufficient condition that the distribution of X_k tend to a non-degenerate limit independent of the distribution of X_0, is that the distribution of ξ have a finite negative mean.

For a homogeneous process modified by two retaining boundaries confining the process to a finite interval, a limiting non-degenerate distribution may be expected if the increments assume both positive and negative values. To see this we note that for such a process on a finite lattice, $n = 0,1,\ldots L$, the states $n = 0$ and the states $n = L$ are both persistent (recurrent), and periodicity of the process is ruled out by the assumption that increments assume both positive and negative values. A discussion of ergodicity for Markov chains may be found in Feller (1957). For a process on the continuum a similar behaviour may be inferred, since the process may be approximated to any desired accuracy by a lattice process.

The replacement process on the half-line $[0,\infty)$ of Section V.1 will also be ergodic if the mean increment μ is negative and if the time between replacements has a finite mean. (When the replacement distribution has a first moment, the latter condition is assured.) For under these conditions the replacement events constitute a class of regeneration events (Smith, 1955a) for the process, and a finite mean time between regenerations implies ergodicity. The replacement process on a finite interval $(0,L)$ will always be ergodic by the same reasoning.

VI.2 The ergodic distribution on the half-line when increments of at least one sign have exponential distribution

An increment distribution $A(x)$ gives rise to a simple ergodic distribution if its negative increments, positive increments, or both negative and positive increments are exponentially distributed. In this section we will discuss the ergodic distribution for the replacement process (V.1.5) in discrete time on the half-line $[0,\infty)$, governed by such an increment distribution $A(x)$ and a replacement distribution $B(x)$. The waiting-time process of Lindley (Example 1 of Section V.1) is a subcase with $B(x) = U(x)$. As we have seen, the replacement process will be ergodic if $\mu = \int x\,dA(x) < 0$ and if $B(x)$ has a finite first moment.

For clarity let us assume initially that $F_0(x)$, $A(x)$ and $B(x)$ are absolutely continuous. The ergodic compensation function $C_\infty(x)$ (Section V.5) and ergodic distribution $F_\infty(x)$ will then be absolutely continuous, and we have from (V.5.5) and (V.5.6)

$$f_\infty(x) = \int_{-\infty}^{\infty} c_\infty(x')\, g(x-x')\, dx' \tag{2.1}$$

$$= c_\infty(x) + \int_{-\infty}^{\infty} c_\infty(x')\, h(x-x')\, dx',$$

where $g(x)$ is the ergodic Green density. From (V.2.12), the compensation density is given by

$$c_\infty(x) = -U(-x) \int_0^\infty a(x-x')\, f_\infty(x')\, dx'$$

$$+ \left\{ \int_{-\infty}^{0} dx \int_0^\infty a(x-x')\, f_\infty(x')\, dx' \right\} b(x). \tag{2.2}$$

There are two cases to be examined, each of distinct character.

Case A. The increments which are exponentially distributed are of the same sign as the mean increment μ

We have assumed $\mu < 0$. Hence we deal for this case with $a(x) = q\eta\, e^{\eta x}\, U(-x) + p\, a_+(x)$, where $a_+(x)$ is a probability density of positive support, and $0 < q < 1$. From $g_A(x) = \delta(x) + h_A(x)$ and Theorem V.8.1, we have

$$g_A(x) = \sum_1^\infty a^{(k)}(x), \qquad x > 0 \tag{2.3a}$$

$$g_A(x) = |\mu|^{-1} \qquad x < 0. \tag{2.3b}$$

From (2.2),

$$c_{A\infty}(x) = \left\{ \int_0^\infty e^{-\eta x'} f_{A\infty}(x')\, dx' \right\} \{-qU(-x)\eta\, e^{\eta x} + qb(x)\}. \tag{2.4}$$

We thus see that the compensation density is known, apart from a multiplicative constant. Hence from (2.1) the ergodic density is known, apart from the constant, and this may be determined from the requirement that $\int f_{A\infty}(x)\, dx = 1$.

When we do not have absolute continuity, (2.4) is replaced from (V.5.2) and (V.1.6b) by $C_{A\infty}(x) = \int e^{-\eta x} dF_{A\infty}(x) \{-q\, e^{\eta x} U(-x) + q\, B(x) - q\, U(x)\}$, and (V.5.8) must be employed in place of (2.1). Of greatest interest is the replacement distribution $B(x) = U(x)$ corresponding to the retaining boundary. We then have

$$C_{A\infty}(x) = -K_A\, e^{\eta x}\, U(-x). \tag{2.5}$$

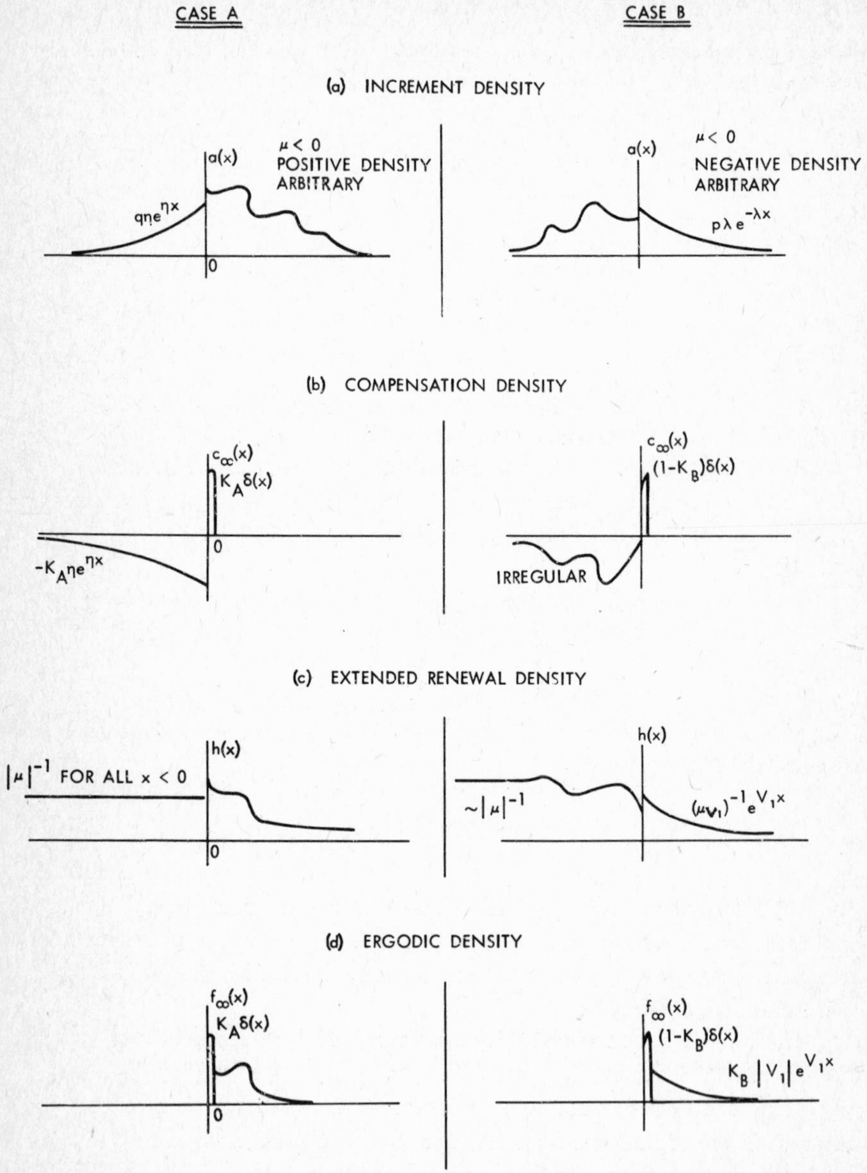

Fig. VI.1—Ergodic distributions for the homogeneous processes on $[0,\infty)$; increments of one sign exponentially distributed; retaining boundary at $x = 0$

When (V.5.8″) is applicable,

$$F_{A\infty}(x) = 1 + H_A^-(x)\cdot C_{A\infty}(x) + C_{A\infty}(x), \tag{2.5'}$$

where $H_A^-(x) = \sum_1^\infty \{A^{(k)}(x) - 1\}$. The positive constant K_A may be identified most quickly from the complex plane. $C_{A\infty}(x)$ has the Fourier-Stieltjes transform $\psi_A(z) = K_A\, iz(\eta + iz)^{-1}$, and we have for the Fourier-Stieltjes transform of $F_{A\infty}(x)$, from V.11.5,

$$\phi_{A\infty}(z) = \frac{K_A\, iz}{\eta + iz}\,\frac{1}{1-\alpha(z)}. \tag{2.6}$$

This characteristic function has the value 1 at $z = 0$, and L'Hôpital's rule gives $K_A = \eta|\mu|$. When $A(x)$ has an extended renewal density $h(x)$, the generalized ergodic density for the retaining boundary may be written via (2.6) in the form

$$f_{A\infty}(x) = \eta|\mu|\left\{\delta(x) + h_A(x) - \int_{-\infty}^{0} \eta e^{\eta x'} h_A(x-x')\,dx'\right\}. \tag{2.7}$$

Case B. The increments which are exponentially distributed have sign opposite to the mean increment μ

Suppose first that $A(x)$ is absolutely continuous. For this case, $a(x) = q\,a_-(x) + p\lambda e^{-\lambda x} U(x)$, where $a_-(x)$ is a probability density of negative support and $0 < q < 1$. Here, by virtue of Theorem V.8.2,

$$\mathbf{g}_B(x) = \mu_{V_1}^{-1} e^{V_1 x} \qquad x > 0 \tag{2.8a}$$

$$\mathbf{g}_B(x) = \sum_1^\infty a^{(k)}(x) \qquad x < 0, \tag{2.8b}$$

where iV_1 is the unique zero of $\{1-\alpha(z)\}$ in the lower half-plane, and μ_{V_1} is the first moment of the conjugate distribution there. Again $f_{B\infty}(x)$ is obtained from (2.1) and $c_{B\infty}(x)$ from (2.2). The form of $c_{B\infty}(x)$ is not simple. Simplicity of form for Case B now resides in $\mathbf{g}_B(x)$ and is transferred through it to $f_{B\infty}(x)$. We have specifically from (2.1) that $f_{B\infty}(x) = \int \mathbf{g}_B(x-x')\, c_{B\infty}(x')\,dx'$. For the retaining boundary case $b(x) = \delta(x)$ and $c_{B\infty}(x) = 0$ for $x > 0$. Hence $f_{B\infty}(x)$ is exponential for $x > 0$ and we infer that the ergodic distribution has the form

$$F_{B\infty}(x) = \{1 - K_B e^{V_1 x}\} U(x). \tag{2.9}$$

It is clear that (2.9) will be valid whether or not $A(x)$ is absolutely continuous.

Again the constant K_B is obtained in the most direct manner from (V.11.5). We note that since $C_{B\infty}(x)$ has no positive support, $\psi_{B\infty}(z)$ is regular in the lower half-plane. Moreover $1 - \alpha(z)$ is analytic for $z \neq -i\lambda$ in the lower half-plane, and $\phi_{B\infty}(z)$ is analytic at all values of z other than $z = iV_1$. Because $\alpha(z)$ has a pole at $z = -i\lambda$, we require that $\phi_{B\infty}(-i\lambda) = 0$. But $\phi_{B\infty}(z) = 1 - K_B + K_B V_1(V_1 + iz)^{-1}$. Hence

$$K_B = 1 - |V_1| \lambda^{-1}. \tag{2.9'}$$

A graphic comparison of Case A and Case B for a retaining boundary at $x = 0$ is given in Fig. VI.1. Some applications follow.

(1) *Lindley waiting-time process: Pollaczek case*

In the theory of queues, the ergodic distributions of this section play a prominent role. They arise in the study of the equilibrium statistics for the Lindley waiting-time process described in Section V.1. Here one has a homogeneous continuum process $X_{k+1} = \max[0, X_k + \xi_{k+1}]$ modified by a single retaining boundary at zero. The random increment $\xi_k = S_k - T_k$ is the difference between two independent random variables, the first governed by a service-time distribution $S(x)$ and the second by an inter-arrival-time distribution, $T(x)$. The increment distribution $A(x)$ of ξ_k has the characteristic function $\alpha(z) = \sigma(z)\,\overline{\tau}(z)$, where $\sigma(z)$ and $\tau(z)$ are the characteristic functions of $S(x)$ and $T(x)$ respectively. When arrivals are independent and T_k has an exponential distribution $E_\eta^+(x) = (1 - e^{-\eta x})\,U(x)$, then $A(x)$ may be seen to have its negative increments of distribution $E_\eta^-(x) = e^{\eta x}$, $x < 0$; $= 1$, $x \geqslant 0$. One then has Case A with the increment distribution given by the convolution

$$A(x) = E_\eta^-(x) \cdot S(x). \tag{2.10}$$

In waiting-time theory, this case is associated with Pollaczek, and the corresponding distribution $F_\infty(x)$ for the particular increment distribution (2.10) is called the Pollaczek† distribution. A modified form of (2.7) for the ergodic distribution is then available as follows. From (2.6) we may write

---

† This distribution is also known as the Khinchin-Pollaczek distribution.

$$\phi_\infty(z) = \frac{K}{1 - \eta\left\{\dfrac{1-\sigma(z)}{-iz}\right\}}. \tag{2.11}$$

We may now identify $\{1-\sigma(z)\}(-iz\bar{S})^{-1}$ as the Fourier transform of the probability density $d(x) = \{1 - S(x)\}/\bar{S}$, where $\bar{S}$ is the mean service time, i.e., the first moment of $S(x)$. For the constant K we have $K = \eta|\mu| = \eta|\bar{S} - \eta^{-1}| = 1 - \eta\bar{S}$. The process will be ergodic when $0 < \eta\bar{S} < 1$. We then have from (2.11), for the ergodic density in the Pollaczek case, a result † given by Kendall (1957):

$$f_\infty(x) = (1-\eta\bar{S})\left\{\delta(x) + \sum_{1}^{\infty}(\eta\bar{S})^n\, d^{(n)}(x)\right\}. \tag{2.12}$$

We see that the Pollaczek distribution is absolutely continuous in every interval not containing $x = 0$.

(2) *Lindley waiting-time process: Smith case*

In the same waiting-time context, the inter-arrival times T_k may have a general distribution $T(x)$, and the service times S_k have an exponential distribution $E_\lambda^+(x)$. We then find Case B and the equilibrium waiting-time distribution has the simple form (2.9). This case is associated with W.L. Smith (1953) who demonstrated a similar simplicity of structure for the waiting-time distribution whenever the service-time distribution has an Erlangian form.

(3) *An inventory process with replacement*

Consider the inventory process described in Section V, Example 3, for which the warehouse has infinite capacity. When the inventory level is depleted, the level is replaced immediately at the value x_R. Let the effective increment distribution $A(x)$ have exponentially distributed negative increments. If $\mu = \int x\, dA(x) < 0$, and the convergence strip for $\alpha(z)$ has a lower boundary $V_{\min} < 0$, then $H^-(x) = \sum_1^\infty\{A^{(k)}(x) - 1\}$ converges and (V.5.8″) is valid. We then have

$$C_\infty(x) = K\{-e^{\eta x}\, U(-x) - U(x) + U(x-x_R)\} \tag{2.13}$$

and

---

† A probabilistic interpretation of (2.12) has been given by Runnenburg (1960).

$$F_\infty(x) = 1 - KK_A^{-1} + KK_A^{-1}F_{A\infty}(x) + K\,\mathbf{G}^-(x-x_R) - K\,\mathbf{G}^-(x), \qquad (2.14)$$

where K_A and $F_{A\infty}(x)$ are given above and $\mathbf{G}^-(x) = U(x) + H^-(x)$. From $\phi_\infty(z) = K\{e^{izxR} - \eta(n+iz)^{-1}\}\{1-\alpha(z)\}^{-1}$ and $\phi_\infty(0) = 1$, we find that $K = |\mu|(x_R + \eta^{-1})^{-1}$.

A feeling for the relative simplicity of this Green's function approach may be obtained from a comparison with the treatment of this process as an inhomogeneous Wiener-Hopf problem (V.12).

VI.3 Processes on the finite interval; two retaining boundaries

Consider now a homogeneous process modified by two retaining boundaries, with single-step transition distribution $A(x',x)$ given by (V.1.6). The Case A of the previous section with a second boundary at $x = L$ is basically no more complex than for a single retaining boundary, as we will see. Let the governing increment distribution $A(x)$ be as described in Theorem V.8.1, with $A(x) = q\,e^{\eta x}$ for negative x and with a negative mean μ for all increments. Then since $C_\infty(x) = 0$ for $0 \leqslant x < L$, (2.1′) will be valid for these values. The compensation function $C_\infty(x)$ will consist of two segments, as shown in Fig. VI.2.

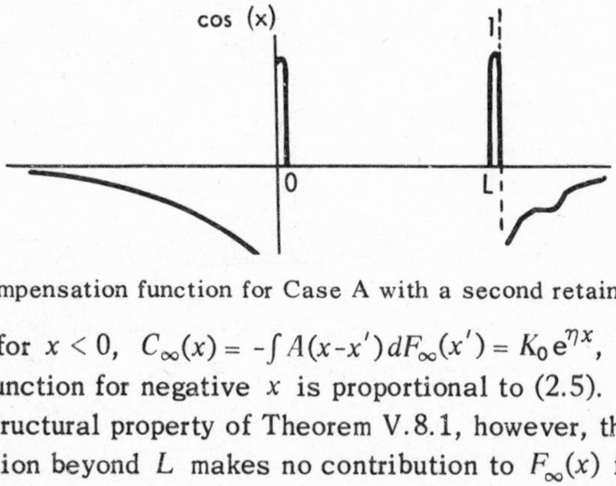

Fig. VI.2–Compensation function for Case A with a second retaining boundary

As before, for $x < 0$, $C_\infty(x) = -\int A(x-x')\,dF_\infty(x') = K_0 e^{\eta x}$, and the compensating function for negative x is proportional to (2.5). By virtue of the basic structural property of Theorem V.8.1, however, the compensation function beyond L makes no contribution to $F_\infty(x)$ for $x < L$. This is most easily seen when $A(x)$ is absolutely continuous via the differential relation (V.5.3). This states that the local ergodic density is

$$f_\infty(x) = \int_{-\infty}^{0+} h(x-x')\,dC_\infty(x') + \int_{L-}^{\infty} h(x-x')\,dC_\infty(x') \quad \text{for } 0 < x < L.$$

But by Theorem V.8.1, $h(x) = |\mu|^{-1}$ for all $x < 0$. Since $\int_{L-}^{\infty} dC_\infty(x')$

$= 0$, the second integral on the right makes no contribution. At $x = L$, $F_\infty(x)$ will be discontinuous with $F_\infty(x) = 1$ for $L \leqslant x$. Hence

$$F_\infty(x) = a_L F_{A\infty}(x) \quad \text{for } x < L$$
$$F_\infty(x) = 1, \qquad L \leqslant x, \tag{3.1}$$

where $F_{A\infty}(x)$ is the distribution (2.5′). When $L = \infty$, $a_L = 1$. The constant a_L may be determined from the requirement that $P\{X_\infty = 0\} = P\{X_\infty + \xi \leqslant 0\}$. If $A(x)$ has no support at $x = 0$, $P\{X_\infty + \xi \leqslant 0\} = q\int_{0-}^{L+} e^{-\eta x} dF_\infty(x)$, i.e.,

$$a_L F_{A\infty}(0) = q\, a_L \int_{0-}^{L} e^{-\eta x}\, dF_{A\infty}(x) + q\, e^{-\eta L}\{1 - a_L F_{A\infty}(L)\}. \tag{3.2}$$

Case B modified by a second retaining boundary at $x = L$, may be treated in a similar manner, and is only slightly more complex. The details will not be given.

VI.4 Lattice processes

The solutions obtained in Sections 2 and 3 have direct analogues for the lattice. Suppose one has the process $N_{k+1} = \max[0, N_k + \xi_{k+1}]$ with a single retaining barrier at $n = 0$, for which the single-step transition probabilities are given by (V.4.2). If $N_0 > 0$ and the increment probabilities are given by $e_n = P\{\xi = n\}$, Lindley's condition for ergodicity (Theorem VI.1.1) is that $\mu = \Sigma\, n e_n < 0$.

The simplest lattice processes are those for which increments in either direction (or both), have a geometric distribution. Unit increments in one direction (or both) are a limiting case. The lattice analogues of Case A and Case B for the continuum have geometrically distributed increments of the same or opposite sign as μ, respectively. These will again be referred to as Case A and Case B.

Case A. The increments which are geometrically distributed are of the same sign as the mean increment

Let the increment probabilities e_n be as described in Theorem V.8.3, i.e., let the increments have a negative mean $\mu = \epsilon'(1) < 0$ and let $e_n = q(1-\beta)\beta^{-n-1}$ for $n < 0$. Consider the homogeneous process on the lattice in discrete time governed by these increment probabilities, modified by a single retaining barrier at $n = 0$. From (V.4.4) and (V.4.2) we find that $d_{n'n} = -e_{n-n'}$ for $n < 0$. Hence, from (V.7.2), we have for $n < 0$ $c_n^{(\infty)} = -\sum_0^\infty p_{n'}^{(\infty)} \mathrm{e}_{n-n'} = -\mathrm{e}_n\{\sum_0^\infty p_{n'}^{(\infty)} \beta^{n'}\} = -K(1-\beta)\beta^{-n-1}$ so that, for $n < 0$, $c_n^{(\infty)}$ has the same geometric dependence on n as e_n. The total compensation $\sum_{-\infty}^{\infty} c_n^{(\infty)}$ must vanish, and this implies that $c_0^{(\infty)} = +K$. From (V.7.3) and (V.7.5) we have therefore $p_n^{(\infty)} = \sum \mathbf{g}_{n-n'} c_{n'}^{(\infty)}$. Hence for $n > 0$,

$$p_n^{(\infty)} = K\delta_{n,0} + Kh_n - K(1-\beta)\beta^{-1}\sum_{-\infty}^{-1} h_{n-n'}\beta^{-n'} \tag{4.1}$$

where h_n is the extended renewal mass (V.7.4). The constant K may again be determined from the complex plane. From (V.11.14) and (V.11.11) we have

$$\pi_\infty(u) = K\,\frac{(u-1)/(u-\beta)}{1-\epsilon(u)}\,. \tag{4.2}$$

From $\pi_\infty(1) = 1$ and L'Hôpital's rule, we then find $K = (1-\beta)|\mu|$. Of particular importance is the case $\beta = 0$, corresponding to unit steps in the negative direction. One then has $p_n^{(\infty)} = K(\mathbf{g}_n - \mathbf{g}_{n+1})$, and since $K = |\mu|$,†

$$p_n^{(\infty)} = |\mu|(\delta_{n,0} + h_n - h_{n+1})\,. \tag{4.3}$$

A second representation of (4.2) is available when $\beta = 0$. We then have $\epsilon(u) = qu^{-1} + p\epsilon_+(u)$, and (4.2) takes the form

$$\pi_\infty(u) = \frac{|\mu|}{q}\,\frac{1}{1-\dfrac{p}{q}\left\{\dfrac{1-\epsilon_+(u)}{1-u}\right\}u}\,.$$

If we set $\theta = (p/q)\mu_+$ where $\mu_+ = \sum_1^\infty n\mathrm{e}_{+n}$, and $d_n = (\mu_+)^{-1}\sum_n^\infty \mathrm{e}_{+j}$ and if we identify $\{d_n : n = 1,2\ldots\}$ as a probability mass distribution on the positive lattice, we have

$$p_n^{(\infty)} = \frac{|\mu|}{q}\sum_0^\infty \theta^j d_n^{(j)}\,. \tag{4.3'}$$

When for Case A a second retaining boundary is present at $n = L$, the discussion runs parallel to that of Section 3 for the analogous case on the continuum. The compensation structure for $n \leqslant 0$ is, apart from

† A result equivalent to (4.3) is given by P.D. Finch (1960), Theorem 5.

some multiplicative constant, identical with that of (4.1). The compensation structure for $n \geqslant L$ is complex in form, but this complexity causes no difficulty, since by Theorem V.8.3 it can contribute only to the state $n = L$. We have therefore

$$p_n^{(\infty)} = K_1 p_{An}^{(\infty)} + K_2 \delta_{n,L},$$

where $p_{An}^{(\infty)}$ is the distribution for $L = \infty$ obtained from (4.2). The coefficients may be found from the normalization condition $\sum_0^L p_n^{(\infty)} = 1$ and the continuity equation $p_0^{(\infty)} = \sum_{n=-\infty}^{0} \sum_{n'=0}^{L} \epsilon_{n-n'} p_{n'}^{(\infty)}$.

The solution for the case $\beta = 0$ maintains its simplicity. By virtue of Theorem V.8.3′, $K_2 = 0$. Hence from (4.3) and $\sum_0^L (g_n - g_{n+1}) = g_0 - g_{L+1}$, we have the concise result (Keilson, 1964b),

$$p_n^{(\infty)} = \frac{g_n - g_{n+1}}{g_0 - g_{L+1}}, \qquad 0 \leqslant n \leqslant L. \tag{4.4}$$

Simple replacement processes may clearly be treated as in Section 2. For the cumulative probabilities we then obtain (I.9.10) with the aid of Theorem V.8.3′.

Case B. The increments which are geometrically distributed are of sign opposite to the mean increment

Suppose now that $\mu < 0$, and that the positive increments have geometric distribution, i.e., let $\epsilon(u) = q\,\epsilon^-(u) + p(1-\alpha)u/(1-\alpha u)$. For this case $\{c_n : n = 0,-1,-2\ldots\}$ is complex, but by virtue of Theorem V.8.4, the ergodic Green masses h_n for $n > 0$, and hence the ergodic state probabilities $p_n^{(\infty)}$, will be geometric in character. Then we may write

$$\pi_\infty(u) = (1-K) + \frac{K(\rho_1 - 1)u}{\rho_1 - u}, \tag{4.5}$$

where ρ_1 is the unique zero of $1 - \epsilon(u)$ on the real axis of magnitude greater than one. The constant K is determined from the requirement that $\pi_\infty(\alpha^{-1}) = 0$ (cf. the discussion justifying (2.9′)). Hence

$$K = \frac{1 - \alpha\rho_1}{(1-\alpha)\rho_1} \tag{4.6}$$

and

$$p_0^{(\infty)} = 1-K; \quad p_n^{(\infty)} = K(\rho_1 - 1)\rho_1^{-n}, \qquad n > 0. \tag{4.7}$$

When two retaining boundaries are present at $x = 0$ and $x = L$, the ergodic distribution for Case B maintains its simplicity. It will be shown in Section VII.7 that for single steps in the negative direction, and positive μ,

$$p_n^{(\infty)} = \frac{\mathbf{g}_n - \mathbf{g}_{n+1} + |\mu_{\rho_1}|^{-1} \rho_1^{-n-1}(1-\rho_1)}{|\mu_{\rho_1}|^{-1} \rho_1^{-L-1} - \mathbf{g}_{L+1}}, \qquad 0 \leqslant n \leqslant L, \tag{4.8}$$

where ρ_1 is the unique zero of $1 - \epsilon(u)$ inside the annulus of convergence with $0 < \rho_1 < 1$.

VI.5 The drift process with increments and drift velocity of opposite sign modified by retaining boundaries

Before discussing the ergodic distributions with two retaining boundaries, it will be convenient to discuss the following replacement process, a version of which is described in Example 5a of Section V.1.

The homogeneous drift process $X(t) = \sum_1^{K(t)} \xi_i + vt$ with $v < 0$ is modified by replacement (Section V.1) at the value T whenever the state $x = 0$ is reached. $K(t)$ is a Poisson process of frequency ν, and the ξ_i are positive independent random variables of distribution $A(x)$. Ergodicity requires that $v + \nu\mu < 0$.

In Section V.3 we saw that the replacement process gives rise to the compensation function $C(x,t) = |v|\, f(0+,t)\{-U(x) + B(x)\}$, when the local density $f(0+,t)$ is meaningful. We also saw in Theorem V.9.1 that the ergodic Green measure $\mathbf{G}_c(x)$ for a homogeneous drift process is absolutely continuous. We therefore anticipate that the ergodic distribution will be absolutely continuous and that even though $f(x,t)$ may be meaningful only in a generalized sense, $f(x,\infty)$ and $f(0+,\infty)$ will be ordinary functions. For replacement at T, $B(x) = U(x-T)$ and we will then have for some $K_T > 0$,

$$f_T(x,\infty) = K_T\{\mathbf{g}_c(x-T) - \mathbf{g}_c(x)\}. \tag{5.1}$$

From (5.1) we note that, since $f_T(x,\infty) > 0$ for all $x > 0$ and $T > 0$, $\mathbf{g}(x)$ is a monotonic decreasing function. The constant K_T is obtained from the complex plane. From (V.11.10) we have $\phi_T(z,\infty) = K_T(e^{izT} - 1) \div \{\nu - \nu\alpha(z) - izv\}$. From L'Hôpital's rule and $\phi_T(0,\infty) = 1$, it follows that $K_T = T^{-1}|\nu\mu + v|$. Note that (5.1) is valid for x negative since the right-hand side reduces to zero by Theorem V.9.2.

Now suppose we consider a sequence of such replacement processes with $T = T_j$ getting infinitesimally small. The compensation function ceases to have meaning in any ordinary sense since K_T becomes infinite. The ergodic density (5.1), however, has for $x > 0$ a well-defined local limit as $T \to 0$, given by

$$f(x,\infty) \;=\; -|\nu\mu + v| \frac{d}{dx} \mathbf{g}_C(x)\,. \tag{5.2}$$

We know that $F(+\infty,\infty) = 1$ and that $\mathbf{g}_C(+\infty) = 0$, and we conclude from (5.2) that

$$F(x,\infty) \;=\; \{1 - |\nu\mu + v|\, \mathbf{g}_C(x)\}\, U(x)\,. \tag{5.3}$$

The limiting process for $T = 0$ corresponds to a retaining boundary at $x = 0$. For when T is some small positive number ϵ, process samples are confined to the interval $(0,\epsilon)$ until a positive increment is experienced, and the retaining character is evident. At $x = 0$, a discontinuity in the ergodic distribution may be anticipated. This may be attributed to the discontinuity in $\mathbf{g}_C(x)$ at $x = 0$, which in turn will always be present when there is a non-vanishing drift velocity v. The discontinuity in $F(x,\infty)$ at $x = 0$ is obtained quickly from $\phi(z,\infty) = \lim_{T\to 0} \phi_T(z,\infty) = |\nu\mu + v|\, iz\{\nu - \nu\alpha(z) - izv\}^{-1}$. From the behaviour of $\phi(z,\infty)$ at $z = \infty$ we infer that $F(x,\infty)$ has a discontinuity $F(0,\infty) = |\nu\mu + v| / |v|$.

An alternative representation is available for the ergodic distribution. The characteristic function $\phi(z,\infty)$ may be rewritten as

$$\phi(z,\infty) \;=\; \left|\frac{\nu\mu + v}{v}\right| \frac{1}{1 - \theta\left\{\dfrac{1 - \alpha(z)}{-iz\mu}\right\}}$$

where $\theta = \nu\mu|v|^{-1}$ and, as for (2.11), we may identify $\{1-\alpha(z)\}(-iz\mu)^{-1}$ as the Fourier transform of the probability density $d(x) = \mu^{-1}\{1 - A(x)\}$. Hence

$$f(x,\infty) = \left|\frac{\nu\mu + v}{v}\right| \left\{ \delta(x) + \sum_{1}^{\infty} \theta^n d^{(n)}(x) \right\}. \tag{5.3'}$$

By virtue of the ergodicity condition $\nu\mu + v < 0$, we have $0 < \theta < 1$ and the convergence of (5.3′) is apparent. Because of Theorem V.6.4 and the positive support of $A(x)$ and $D(x)$, the series in (5.3′) will converge for all values of θ. This will be used shortly in the discussion of the same process modified by a second retaining boundary.

If the drift velocity v is positive and the increments are negative, we anticipate that the ergodic distribution of $X(t) = vt + \sum_1^{K(t)} \xi_i$ for a retaining boundary at zero will be exponential. For one has, for any jump process $X_j(t)$ (II.7) approaching the drift process, exponentially distributed positive increments. Hence from Case B of Section 2, every ergodic distribution $F_j(x,\infty)$ is exponential and $F(x,\infty)$ will be exponential. (The same structure may be predicted from Theorem V.9.3, since for any compensation function $C(x)$ with non-positive support, $\int_{-\infty}^{0+} g(x - x')\,dC(x')$ will be exponential for $x > 0$.) For the jump process $X_j(t)$ we have, from (2.9′), $K_j = 1 + V_{1j}\,\lambda_j^{-1}$. The sequence V_{1j} approaches the finite $V_1 < 0$ found as in Theorem V.9.3. The sequence $\lambda_j \to \infty$. Hence $K = \lim K_j = 1$, and the ergodic distribution $F(x,\infty)$ is absolutely continuous with density

$$f_\infty(x) = |V_1|\, e^{-|V_1|x}\, U(x). \tag{5.4}$$

The absence of a probability mass at $x = 0$ may be understood intuitively, since the positive drift velocity acts to prevent any accumulation there.

Consider next the drift process modified by two retaining boundaries at $x = 0$ and $x = L$ with negative velocity, positive increments, and negative mean increment per unit time $\nu\mu + v < 0$. A process of this kind, we have seen, provides a model for a reservoir with finite capacity and uniform release rate. To discuss the ergodic distribution we may employ the same sequence of jump processes $X_j(t)$ with negative increments exponentially distributed to approximate $X(t)$. By virtue of Theorem V.9.2 the compensation function with support on (L,∞) will not contribute to $F_j(x,\infty)$ for $x < L$ for any of the processes $X_j(t)$. The discontinuity in $F_j(x,\infty)$ at $x = L$ vanishes as $j \to \infty$, because of the negative drift velocity. Thus we obtain, from (5.3), $F(x,\infty) = K\{1 - |\nu\mu + v|\, \mathbf{g}_c(x)\}$. The normalization condition $F(L,\infty) = 1$, gives the final result

$$F(x,\infty) = \frac{1 - |\nu\mu + v|\,\mathbf{g}_c(x)}{1 - |\nu\mu + v|\,\mathbf{g}_c(L)}\,U(x), \quad x \leqslant L; \qquad F(x,\infty) = 1, \quad x > L, \tag{5.5}$$

It must be noted that the ergodicity condition appropriate to the single retaining barrier at $x = 0$ is of no interest when a second retaining barrier is present. For $v < 0$, the process is ergodic for all values of $\theta = \nu\mu|v|^{-1} \neq 0$. The ergodic distribution has the characteristic function $\phi(z,\infty) = \{Kiz + \psi_2(z)\}\{\nu - \nu\alpha(z) - izv\}^{-1}$, where $\psi_2(z)$ is the transform of the compensation function component on $[L,\infty)$. The representation (5.3′) is again valid for the inverse transform of $Kiz\{\nu - \nu\alpha(z) - izv\}^{-1}$, the only part of $\phi(z,\infty)$ which makes a contribution. Hence $F(x,\infty)$ has† the alternative representation for $0 \leqslant x \leqslant L$,

$$F(x,\infty) = \frac{U(x) + \sum_1^\infty \theta^n D^{(n)}(x)}{1 + \sum_1^\infty \theta^n D^{(n)}(L)}. \tag{5.5′}$$

The same reasoning carries through to the more general skip-free process on the continuum $X(t) = X_J(t) + vt + X_D(t)$ where the drift rate v and increments for the jump process are again of opposite sign. In Theorem V.9.4 we saw that the ergodic Green density $\mathbf{g}_c(x)$ is convergent and continuous for all x when $D > 0$ and $E\{X(1)\} = \mu(1) < 0$. Moreover, from Theorem V.9.5, if $v < 0$ and the increments are positive, $\mathbf{g}_c(x) = |\mu(1)|^{-1}$ for all $x \leqslant 0$. As for (5.2), we quickly conclude that $f(x,\infty) = -K(d/dx)\,\mathbf{g}_c(x)$, and hence that

$$F(x,\infty) = \frac{\mathbf{g}_c(0) - \mathbf{g}_c(x)}{\mathbf{g}_c(0)} = 1 - |\mu(1)|\,\mathbf{g}_c(x). \tag{5.6}$$

For two retaining boundaries, we obtain, for $0 \leqslant x \leqslant L$,

$$F(x,\infty) = \frac{1 - |\mu(1)|\,\mathbf{g}_c(x)}{1 - |\mu(1)|\,\mathbf{g}_c(L)}. \tag{5.7}$$

Since for $D = 0$, $\mu(1) = v + \nu\mu$, (5.6) and (5.7) pass continuously to

† Gaver and Miller (1962) have obtained this result by Laplace transformation of the continuity equation. Ruin probabilities for the process are derived in the same way by Keilson (1963c).

the previous results for the drift processes. Note however that for $D = 0$, the discontinuity at $x = 0$ in $g_C(x)$ gives rise to a mass concentration at $x = 0$.

When $D \neq 0$, $v > 0$, increments are negative, $\mu(1) < 0$, and we again find (5.6).

VI.5.1 The presence of a double layer

The following remarks may be omitted without loss of continuity.

The character of the result (5.2) is very similar to that found in potential theory under related circumstances. One has in effect in (5.2) the equivalent of a double distribution or magnetic shell at $x = 0$.† The potential of the double distribution is the derivative of the ordinary potential $g_C(x)$. The appearance of a double layer at a reflecting boundary is seen most clearly for diffusion in three dimensions. Suppose one has diffusion in a finite domain D with surface S. A subset S_A of S is designated as absorbing, and replacement at r_0 is specified when S_A is reached. The complementary surface $S_R = S - S_A$ is assumed to be reflecting. It may then be seen that the density $f(r,t)$ satisfies the equation

$$\partial f/\partial t = D\nabla^2 f + I(t)\,\delta(r - r_0) \tag{5.8}$$

where $I(t)$ is the extended renewal density for arrivals at the absorbing surface. The boundary conditions are that $f(r,t) = 0$ at S_A, and $\frac{\partial}{\partial n} f(r,t) = 0$ at S_R. As $t \to \infty$, an ergodic distribution is reached, and $f(r,\infty)$ satisfies the equation $D\nabla^2 f + I(\infty)\,\delta(r - r_0) = 0$. The solution of this equation, in the absence of boundaries, for $r_0 = 0$ and $I(\infty) = 1$ is

$$f_\infty(r) = g_C(r) = \frac{1}{4\pi D|r|} \tag{5.9}$$

where $g_C(r)$ is the ergodic Green's function. By the use of Green's identity†, it may then be seen that for the required boundary conditions,

$$f_\infty(r) = I(\infty)\,g_C(r - r_0) + \int_{S_A} \frac{\partial}{\partial n} f_\infty(r')\,g_C(r - r')\,d\sigma'$$

$$- \int_{S_R} f_\infty(r') \frac{\partial}{\partial n'} g_C(r - r')\,d\sigma'. \tag{5.10}$$

---

† See, e.g., Kellogg, O.D., *Potential Theory*, Dover, 1953, N.Y.

The first integral on the right corresponds to a single layer at the absorbing surface, the second to a double layer at the reflecting surface. The positive constant $I(\infty)$ may be obtained from normalization.

If the entire surface S is reflecting, and there is no replacement, only the last term on the right-hand side of (5.10) is present, and this is the analogue of (5.2).

VI.6 Asymptotic behaviour of the ergodic distributions

The asymptotic behaviour of the ergodic distributions may be inferred in simple cases from that of the ergodic Green's functions. The theorems of Section V.10 are of particular value in this regard.

Theorem VI.6.1. Let X_k be a homogeneous process on the continuum modified by a single retaining boundary at $x = 0$, for which $X_k = \max[0, X_{k-1} + \xi_k]$. If the increment distribution $A(x)$ satisfies the conditions of Theorem V.10.1, the local density $f_\infty(x)$ of the ergodic distribution has the asymptotic representation †

$$f_\infty(x) \sim \frac{K}{\mu_{V_1}} e^{V_1 x} \quad \text{as } x \to \infty; \qquad 0 < K < 1. \tag{6.1}$$

For from (V.5.3), $f_\infty(x) = \int h(x-x')\,dC_\infty(x')$ when $x > 0$, since the compensation function $C_\infty(x)$ has only non-positive support. From (V.10.2) we then have (6.1), where $K = \int_{-\infty}^{0+} e^{-V_1 x}\,dC_\infty(x)$. This may be seen to have the form $K = C_0[1 - \int_{-\infty}^{0} e^{-V_1 x}\mu(x)\,dx]$ where $C_0 = P\{X_\infty + \xi < 0\}$ is the discontinuity in $C_\infty(x)$ at $x = 0$, and $\mu(x) > 0$ with $\int \mu(x)\,dx = 1$. Hence $0 < K < 1$. Q.E.D.

The analogue of (6.1) for the half-lattice may be inferred from (V.10.4).

Theorem VI.6.2. Let $N_k = \max[0, N_{k-1} + \xi_k]$. If the increment probabilities e_n have a negative mean $\mu = \sum n e_n < 0$, and the generating function $\epsilon(u)$ has an annulus of convergence containing the value $\rho_1 > 1$ at which $\epsilon(\rho_1) = 1$, then the ergodic state probabilities have the asymptotic representation

$$p_n^{(\infty)} \sim \frac{K}{\mu_{\rho_1}} \rho_1^{-n} \quad \text{as } n \to \infty; \qquad 0 < K < 1. \tag{6.2}$$

The reasoning is identical with that for Theorem VI.6.1, and (6.2) follows from (V.7.3) and (V.10.9).

---

† Of related interest is the important limit theorem of Kingman describing the exponential character under certain conditions of the ergodic distribution when the mean increment goes to zero. See Kingman (1965).

As we have seen in (V.5.1), the ergodic distribution for a jump process is identical with that for the discrete time process with which it is associated, and Theorems 6.1 and 6.2 are directly applicable. For a drift process with negative drift rate and positive increments, as described in Section 5, we may employ (5.2) and Theorem V.10.3 to infer that, as $x \to \infty$,

$$f(x,\infty) \sim \frac{\mu(1)\,V_1}{\mu_{V_1}(1)}\, e^{V_1 x}, \tag{6.3}$$

where $\mu(1) = \nu\mu + v$, $\mu_{V_1}(1)$ is given by (V.10.5′), and V_1 is obtained from (V.10.7) with $D = 0$. The constant of proportionality for this case is therefore known exactly. For more general homogeneous processes in continuous time such as (V.10.3), we may again infer that as $x \to \infty$

$$f(x,\infty) \sim \alpha\, e^{V_1 x}. \tag{6.4}$$

A bound or estimate for the coefficient α is not available for the general case.

For a process on a finite interval with two retaining boundaries, the asymptotic behaviour at a sufficient distance from both boundaries may be exhibited in the same way when the convergence strip has a suitable structure. For this approximation to have a useful domain of validity, the length of the interval must be greater than the larger of the distances required by the ergodic Green density to approximate its limiting behaviour in the positive and negative directions. The contributions from the two boundaries are weighted by two coefficients which may be approximated by perturbation methods. Details will not be given.

VI.7 The gambler's ruin problem

The classical gambler's ruin problem (Feller, 1957) treats the homogeneous lattice process N_k in discrete time with Bernoulli increments (Section IV.2). The process commences at a prescribed state $N_0 = m$ on the lattice interval $-1 < n < L$. The probability R_m of reaching the state $n = L$ (ruin) before the state $n = -1$ (success) is sought. In this and the following section, we wish to discuss a more general version of the same problem. One is given a spatially homogeneous process X_k governed by a given increment distribution $A(x)$, and a prescribed initial value y on the half-open interval $[0,L]$; one seeks the *ruin probability* R_y of reaching the set $[L,\infty)$ (ruin) before reaching $(-\infty,0)$ (success).

For clarity, we will first discuss lattice processes. Let N_k be the homogeneous lattice process commencing at m and governed by the increment probabilities $\{e_j\}$. The ruin probability R_m is simply related to the passage-time probability $r_m^{(k)}$ that the process will terminate in failure at the kth step if it starts at $N_0 = m$. For we have for $0 \leqslant m \leqslant L-1$

$$r_m^{(k)} = \sum_{j=0}^{L-1} e_{j-m} r_j^{(k-1)} \qquad \text{for } k \geqslant 2 \tag{7.1}$$

and

$$r_m^{(1)} = \sum_{L-m}^{\infty} e_j . \tag{7.2}$$

If $R_m^{(k)}$ is the probability of ruin at or before the kth step, then $R_m^{(k)} = \sum_{j=1}^{k} r_m^{(j)}$. Hence from (7.1) and (7.2) we have the recursive equations, valid for $k \geqslant 1$,

$$R_m^{(k)} = \sum_{j=0}^{L-1} e_{j-m} R_j^{(k-1)} + \sum_{L-m}^{\infty} e_j , \qquad 0 \leqslant m \leqslant L-1, \tag{7.3}$$

where $R_m^{(0)} = 0$. Since $R_m^{(k)}$ is clearly monotonic increasing in k and bounded by unity, the sequence $\{R_m^{(k)}, k = 1,2,\ldots\}$ has a limit which must be the ruin probability $R_m = \lim_{k\to\infty} R_m^{(k)}$. From (7.3), R_m satisfies the set of equations

$$R_m = \sum_{j=0}^{L-1} e_{j-m} R_j + \sum_{L-m}^{\infty} e_j , \qquad 0 \leqslant m \leqslant L-1 . \tag{7.4}$$

These equations have a unique solution provided by Cramer's Rule, but if L is large this solution may be cumbersome. As we will now show, however, the set of numbers R_m are simply related to the ergodic distribution for an associated process, and we are enabled thereby to apply our results for ergodic distributions to advantage.

Consider the process N_k on the finite lattice $n = 0,1,2,\ldots,L$ having two retaining boundaries at $n = 0$ and $n = L$ and governed by the single-step transition probabilities e_{mn} given by (V.4.3). The equations of continuity governing the state probabilities $p_n^{(k)} = P\{N_k = n\}$ are

$$p_n^{(k)} = \sum_{m=0}^{L} p_m^{(k-1)} \mathrm{e}_{mn}, \qquad k \geqslant 1. \tag{7.5}$$

The corresponding equations governing the cumulative state probabilities $P_n^{(k)} = P\{N_k \leqslant n\} = \sum_{-\infty}^{n} p_m^{(k)}$ are needed. To obtain these from (7.5) we introduce the cumulative transition probabilities $E_{mn} = \sum_{-\infty}^{n} \mathrm{e}_{mn'}$, for which we have from (V.4.2)

$$\begin{aligned} E_{mn} &= 0 && n < 0 \\ E_{mn} &= \sum_{-\infty}^{n-m} \mathrm{e}_j && 0 \leqslant n < L, \quad 0 \leqslant m \leqslant L \\ E_{mn} &= 1 && L \leqslant n. \end{aligned} \tag{7.6}$$

From (7.5) we may write

$$P_n^{(k)} = \sum_{0}^{L} p_m^{(k-1)} E_{mn} = \sum_{0}^{L} \{P_m^{(k-1)} - P_{m-1}^{(k-1)}\} E_{mn}$$

$$= \sum_{0}^{L-1} P_m^{(k-1)} \{E_{mn} - E_{m+1,n}\} + P_L^{(k-1)} E_{Ln}.$$

From $E_{mn} - E_{m+1,n} = \mathrm{e}_{n-m}$ for $0 \leqslant m \leqslant L$, $0 \leqslant n < L$, we have finally for $k \geqslant 1$

$$P_n^{(k)} = \sum_{m=0}^{L-1} P_m^{(k-1)} \mathrm{e}_{n-m} + P_L^{(k-1)} \sum_{-\infty}^{n-L} \mathrm{e}_j, \qquad 0 \leqslant n < L, \; k \geqslant 1. \tag{7.7}$$

Since we are dealing with a probability distribution,

$$P_L^{(k)} = 1. \tag{7.8}$$

It may be observed that (7.7) and the prescribed initial distribution generate $P_n^{(k)}$ by recursion.

We may now compare equations (7.3) and (7.7). If we introduce the adjoint set of increment probabilities $\tilde{\mathrm{e}}_j = \mathrm{e}_{-j}$, we see that (7.3) becomes

$$R_m^{(k)} = \sum_{j=0}^{L-1} \tilde{e}_{m-j}\, R_j^{(k-1)} + \sum_{-\infty}^{m-L} \tilde{e}_j \tag{7.9}$$

and e_j is replaced by $\tilde{e}_j$ in (7.7), *provided that* $P_j^{(0)} = 0$ *for* $0 \leqslant j \leqslant L-1$. Hence we have proved the following theorem:

Theorem VI.7.1. Let N_k be the homogeneous lattice process $N_k = m + \sum_1^k \xi_i$, with increment probabilities $e_j = P\{\xi_i = j\}$. Let $\tilde{N}_k$ be the process adjoint to N_k commencing at L. Finally let N_k' be the process $\tilde{N}_k$ modified by retaining barriers at $n = 0$ and $n = L$. Then the probability $R_m^{(k)}$ of ruin for the process N_k at or before the kth step is identical with the cumulative state probabilities $P_m'^{(k)} = P\{N_k' \leqslant m\}$ for the process N_k', i.e.,

$$R_m^{(k)} = P_m'^{(k)} \qquad \text{for } k \geqslant 1, \quad 0 \leqslant m \leqslant L-1. \tag{7.10}$$

For the probability of ultimate ruin, we have

$$R_m = P_m'^{(\infty)} \qquad 0 \leqslant m \leqslant L-1. \tag{7.10'}$$

Simple ruin distributions for the lattice may now be exhibited with the aid of the results of Section VI.4. The ruin distribution for single steps in one direction and arbitrary step size in the other direction has been given previously by a different method in Keilson (1964c).

Consider now the more general process on the continuum with increment distribution $A(x)$. Since the continuum process may be approximated to any desired accuracy by a lattice process, it is obvious that the structure of the result in Theorem 7.1 will be maintained. Hence we have:

Theorem VI.7.2† For the homogeneous process $X_k = y + \sum_1^k \xi_i$, with increment distribution $A(x) = P\{\xi_i \leqslant x\}$, the probability $R_y^{(k)}$ of ruin at or before the kth step is identical with the distribution $F_k'(x)$ for X_k', the process with adjoint increment distribution $\tilde{A}(x) = 1 - A(-x)$, modified by retaining barriers at $x = 0$ and $x = L$, and commencing at $X_0' = L$; i.e.,

$$R_y^{(k)} = F_k'(y). \tag{7.11}$$

The probability of ultimate ruin is the ergodic distribution

$$R_y = F_\infty'(y) \tag{7.11'}$$

† See Note 11.

For processes in continuous time, the extension of Theorem 7.2 is immediate. For such a process, $X(t)$, commencing at $X_0 = y$, there will be a ruin probability $R_y(t)$ of arriving at the interval $[L,\infty)$ without having visited the interval $(-\infty,0)$ during $(0,t)$. If the process $X(t)$ is a jump process of frequency ν associated with the discrete time process X_k, (II.2), it is clear that

$$R_y(t) = \sum_0^\infty \{(\nu t)^k e^{-\nu t}/k!\} R_y^{(k)}.$$

If $X'(t)$ is the jump process in continuous time of frequency ν associated with the modified process X'_k, its distribution at time t is $F'(x,t) = \Sigma\{(\nu t)^k e^{-\nu t}/k!\} F'_k(x)$. Hence from (7.11) we have the analogue of (7.10) for jump processes in continuous time

$$R_y(t) = F'(y,t). \tag{7.12}$$

The more general homogeneous process (II.3.20), having a Wiener-Lévy component and drift component, may be represented (cf. II.7) as the limit of a sequence of jump processes $X_j(t)$ for each of which (7.12) will be valid. Consequently (7.12) will continue to be valid for the limiting process as well. Thus let $X(t)$ be the homogeneous process of (II.3.20),

$$X(t) = y + X_J(t) + X_D(t) + vt \tag{7.13}$$

where $X_J(t)$ is the jump process commencing at zero, of frequency ν and increment distribution $A(x)$. The adjoint process $\tilde{X}(t)$, defined as the limit of the sequence $X_j(t)$ of adjoint jump processes in the sense of Theorem 7.2, is easily seen to be

$$\tilde{X}(t) = L + \tilde{X}_J(t) + X_D(t) - vt \tag{7.14}$$

where $\tilde{X}_J(t)$ is the adjoint jump process commencing at zero of frequency ν and increment distribution $1 - A(-x)$. Let $X'(t)$ be the process $\tilde{X}(t)$ (which commences at L) modified by retaining boundaries at $x = 0$ and $x = L$. We then have

Theorem VI.7.3. The probability $R_y(t)$, for the process $X(t)$ of (7.13), of ruin at or before time t, is given by

$$R_y(t) = F'(y,t). \tag{7.15}$$

The probability of ultimate ruin is the ergodic distribution of $X'(t)$, i.e.,

$$R_y = F'(y,\infty). \tag{7.15'}$$

For a homogeneous drift process, with negative velocity and positive increments, the probability of reaching $x = 0$ before reaching $[L,\infty)$ from an initial value y is easily seen from (7.15′) to be $F(L-y,\infty)$, where $F(x,\infty)$ is given by (5.5′). This may be seen to be in agreement with (16) of Keilson (1963c). The identification (7.15′) with the ergodic distribution was suggested by a related observation of J. Runnenburg for this case (private communication).

The following identities are convenient. Let N_k be a homogeneous lattice process, $\tilde{N}_k$ be the homogeneous process with adjoint increments, N_k^* be N_k modified by retaining boundaries at 0 and L, and N_k' be $\tilde{N}_k$ modified by the same retaining boundaries. Then clearly $p_m^{*(\infty)} = p_{L-m}'^{(\infty)}$. From this we find that $P_m'^{(\infty)} = 1 - P_{L-m-1}^{*(\infty)}$. Consequently (7.10′) becomes

$$R_m = 1 - P_{L-m-1}^{*(\infty)}, \qquad 0 \leqslant m \leqslant L-1. \tag{7.16}$$

Similarly, for (7.11′) we find

$$R_y = 1 - F_\infty^*(L-y-0), \qquad 0 \leqslant y < L, \tag{7.17}$$

and in place of (7.12),

$$R_y = 1 - F^*(L-y-0,\infty), \qquad 0 \leqslant y < L. \tag{7.18}$$

VI.8 The distribution of extremes for bounded and unbounded processes

One may also consider the same ruin problem with a single boundary by permitting L to become infinite. Ruin then corresponds to the probability of escape to infinity. When the process X_k has a net motion towards the origin with time, the process X_k' is not ergodic and R_y is clearly zero. When the process X_k has a net motion away from the origin with time, X_k' will be ergodic and $R_y = F_\infty'(y)$. The result may be summarized as follows.

Theorem VI.8.1. For the homogeneous process X_k governed by the increment distribution $A(x)$, with positive mean increment $\mu = \int x\, dA(x) > 0$, the probability that X_k will ever reach zero if $X = y > 0$ is given by

$$1 - R_y = 1 - F'_\infty(y). \tag{8.1}$$

The probability that a homogeneous process $X(t)$ with positive mean increment per unit time will ever reach zero if $X(0) = y > 0$ is

$$1 - R_y = 1 - F'(y,\infty). \tag{8.2}$$

The result in Theorem 8.1 may now be employed to give the distribution of extremes for homogeneous processes. For any homogeneous process X_k governed by increments of negative mean, commencing at $X_0 = 0$, every process sample $x_k(\omega)$ will have a finite maximum $e(\omega)$ called the *extreme*. From (8.1), it may be seen that this random variable will have the distribution $E(x)$ given by

$$E(x) = F^*_\infty(x) \tag{8.3}$$

where $F^*_\infty(x)$ is the ergodic distribution for the process X^*_k obtained when X_k is modified by a retaining barrier at zero. For the probability that the maximum exceeds x is the probability that the process $\tilde{X}_k$ commencing at $x > 0$ with adjoint increments will reach zero, and (8.1) may be invoked. Equation (8.3) may be derived directly by the following argument. Let E_k be the maximum value that $X_k = \sum_1^k \xi_i$ will attain. Then it is easy to see, from the situation after a first step, that $E_k = \max[0, \xi_k + E_{k-1}]$ for $k \geqslant 1$. But this is the equation for the process X^*_k, and (8.3) is verified. The analogue of (8.3) for any homogeneous process $X(t)$ is

$$E(x) = F^*(x,\infty), \tag{8.4}$$

where $F^*(x,\infty)$ is the ergodic distribution for the process $X^*(t)$, the modification of $X(t)$ with a retaining boundary at zero. This is a direct consequence of DeFinetti's Theorem, II.7.1. Equations (8.1) and (8.3) are familiar results in random walk theory. The reader is referred to Spitzer (1964, Chapter IV).

Of equal interest is the extreme described in Section I.1 for a homogeneous process modified by a single retaining boundary at $x = 0$. In discrete time the process of interest, X_k, is governed by the law $X_{k+1} = \max[0, X_k + \xi_{k+1}]$, and the random variables ξ_k have the common distribution $A(x)$. When $\mu = \int x\, dA(x) < 0$ and $A(x)$ has some positive support, the process is ergodic and the state $x = 0$ is recurrent. If a positive increment occurs when a process sample is in the state $x = 0$, the sample will assume a sequence of positive values terminating when the sample returns to $x = 0$. Let ζ be the maximum value or extreme taken on by the sample for such a "tour". The extremes for successive tours constitute a sequence of independent random variables, and have some distribution $E(z) = P\{\zeta \leqslant z\}$.

The discussion will again be clearer if first restricted to the lattice. Let ζ_m be the maximum value reached by N_k before returning to 0 if at $k = 0$, $N_0 = m > 0$. Clearly, $P\{\zeta_m \leqslant N\}$ is the probability of reaching the lattice interval $-\infty < n \leqslant 0$ before reaching the lattice interval $N + 1 \leqslant n$. Hence $P\{\zeta_m \leqslant N\}$ is the probability of success for the ruin problem on $[0,N)$ when one commences at $N_0 = m - 1$. From (7.16) we then have

$$P\{\zeta_m \leqslant N\} = P^{*(\infty)}_{N-m}\Big|_{[0,N]}, \qquad 1 \leqslant m, \tag{8.5}$$

where $P^{*(\infty)}_{m}\Big|_{[0,N]}$ is the cumulative probability for the ergodic distribution when N_k is modified by retaining boundaries at $n = 0$ and $n = N$.

For the tour extremes on the lattice, it then follows that

$$P\{\zeta \leqslant N\} = \sum_1^\infty e_m \, P^{*(\infty)}_{N-m}\Big|_{[0,N]} \Big/ \sum_1^\infty e_m . \tag{8.6}$$

Consider, in particular, the ergodic lattice process N_k with unit increments only in the negative direction, arbitrary positive increments, and a single retaining boundary at $n = 0$. We may then employ (8.5) and (4.4) with simple algebra to give for $1 \leqslant m$,

$$P\{\zeta_m \leqslant N\} = \left\{\frac{\mathbf{g}_0 - \mathbf{g}_{N+1-m}}{\mathbf{g}_0 - \mathbf{g}_{N+1}}\right\} U(N - m), \tag{8.7}$$

where $\mathbf{g}_n$ is the ergodic Green mass (V.7.5) for the underlying homogeneous process.†

For ergodic processes on the continuum, the following theorem is obtained from (8.5).

---

† From (8.7) and Theorem V.10.4 an asymptotic representation for $[1-P\{\zeta_m \leqslant N\}]$ may be exhibited at once. By simple algebra, and use of $\mathbf{g}_0 = |\mu|^{-1}$ (Theorem V.8.3′), we find

$$1 - P\{\zeta_m \leqslant N\} \sim \left|\frac{\mu}{\mu_{\rho_1}}\right| \rho_1^{-N-1} (\rho_1^m - 1) \qquad \text{as } N \to \infty . \tag{8.7′}$$

Theorem VI.8.2. Let X_k be the homogeneous process $X_k = y + \sum_1^k \xi_i$, governed by the increment distribution $A(x)$. Let ζ_y be the maximum value attained by X_k before reaching the open set $(-\infty,0)$, and let $E_y(z) = P\{\zeta_y \leqslant z\}$. Then

$$E_y(z) = F_\infty^*(z-y)\Big|_{[0,z]}, \qquad 0 < y \tag{8.8}$$

where $F_\infty^*(x)\Big|_{[0,z]}$ is the ergodic distribution for the homogeneous process X_k when modified by retaining boundaries at $x = 0$ and $x = z$.

A direct extension of the reasoning along the previous lines shows that for any ergodic homogeneous process $X(t)$, modified by a retaining boundary at zero, the largest value ζ_y attained before returning to the state $x = 0$ has the distribution

$$E_y(z) = P\{\zeta_y \leqslant z\} = F^*(z-y,\infty)\Big|_{[0,z]}, \qquad 0 < y, \tag{8.9}$$

where $F^*(x,\infty)\Big|_{[0,z]}$ is the ergodic distribution for $X(t)$ modified by retaining boundaries at $x = 0$ and $x = z$.

For the homogeneous drift process, modified by a single retaining boundary at zero, we then have from (5.5) for $y > 0$,

$$E_y(z) = \frac{1 - |\nu\mu + v|\, g_c(z-y)}{1 - |\nu\mu + v|\, g_c(z)}\, U(z-y). \tag{8.10}$$

The maximum value for a tour is then

$$E(z) = P\{\zeta \leqslant z\} = \int_{0+}^{\infty} E_y(z)\, dA(y) \Big/ \{1 - A(0+)\}. \tag{8.11}$$

CHAPTER VII

PROCESSES WITH REPRESENTATION OF FINITE DIMENSION

For homogeneous processes on the continuum or lattice modified by a single retaining boundary at $x = 0$, a finite representation is available for the description of the process when the increment distributions have a negative or positive component of a certain class. Wiener-Hopf and Hilbert problem solutions are available in principle under the same circumstances, but such solutions are often formal. A similar representation of finite dimension is available for processes on a finite interval or finite lattice. To attain the finite representation, a generalized form of compensation is required.

VII.1 A finite basis for the representation of replacement processes on the half-line

For a certain class of increment distributions $A(x)$, the replacement processes on a finite or semi-infinite interval (Section V.1) may be conveniently described in a linear vector space of finite dimension associated with the compensation function. The availability of such a finite basis for the description is the direct analogue of the factorizability of the kernels arising when the process is characterized by a Wiener-Hopf or Hilbert equation (V.12). Analytical formulations of the Wiener-Hopf and Hilbert type, however, are often frustrated by an inability to locate the zeros of the kernel needed for factorization or to carry out the subsequent inversions required. The finite representation provides an alternative formulation in the real domain, and circumvents factorization difficulties.

In Section V.1 we considered a replacement process X_k on the interval $[0,\infty)$ characterized by an increment distribution $A(x)$ and replacement distribution $B(x)$. Let us consider now the same process with the possibility of termination instead of replacement. When the value $X_k + \xi_{k+1}$ is negative, the process will be replaced with distribution $B(x)$ on the interval $[0,\infty)$ with probability θ, and terminated with probability $(1-\theta)$. The state space $0 \leqslant x < \infty$ may be supplemented by an absorbing state $\mathfrak{U}$. The probability mass at the state $\mathfrak{U}$ represents samples that have terminated. The discussion will be simplified by the assumption that $A(x)$, $B(x)$ and the initial distribution $F_0(x)$ are absolutely continuous, an assumption which may be lifted easily afterwards. Let $f_k(x)$ be the joint probability density that the system has not terminated and that $X_k = x$; and let u_k be the probability that the process X_k is in the state $\mathfrak{U}$. Conservation of probability requires that $u_k + \int_0^\infty f_k(x')\,dx' = 1$. Transitions within the subspace $[0,\infty)$ are described for $k \geqslant 1$ by the continuity equation

$$f_k(x) = \int_0^\infty f_{k-1}(x')\,a(x',x)\,dx' \tag{1.1}$$

where† $a(x',x) = a(x-x')U(x) + \theta A(-x')b(x)$. Hence we may write for $k \geqslant 1$ and all x

$$f_k(x) = \int_0^\infty f_{k-1}(x')a(x-x')\,dx' + c_k(x) \tag{1.2}$$

where†

$$c_k(x) = \theta I_k b(x) - U(-x)\int_0^\infty a^-(x-x')f_{k-1}(x')\,dx', \tag{1.3}$$

and $I_k = \int f_{k-1}(x')A(-x')\,dx'$. In (1.3) we have introduced the notation $f^+(x) = f(x)U(x)$, $f^-(x) = f(x)U(-x)$, and have employed the obvious relation $U(-x)\int_0^\infty a(x-x')f(x')\,dx' = U(-x)\int_0^\infty a^-(x-x')f(x')\,dx'$. We may now derive the recursive relations governing the functions $c_k^-(x)$. The negative component of (1.3) is

$$c_{k+1}^-(x) = -\left\{\int_0^\infty a^-(x-x')f_k(x')\,dx'\right\} U(-x). \tag{1.4}$$

and we note that

$$I_k = -\int_{-\infty}^0 c_k^-(x)\,dx. \tag{1.4'}$$

The functions $f_k(x)$ may be eliminated by employing the analogue of

---

†Note that when $\theta < 1$, $f_k(x)$ is not a true probability density, i.e. $\int f_k(x)\,dx \leqslant 1$. Similarly $\int a(x',x)\,dx \leqslant 1$, and $\int c_k(x)\,dx$ does not have to be zero.

(V.2.3). From (1.2) we infer that $f_k(x) = f_0(x) \cdot a^{(k)}(x) + \sum_1^k c_{k'}(x) \cdot a^{(k-k')}(x)$. Since

$$c_k(x) = \theta I_k b(x) + c_k^-(x) \tag{1.3'}$$

from (1.3), and $f_k(x) = [f_k(x)]^+$, we have for $k \geqslant 1$

$$f_k(x) = [f_0(x) \cdot a^{(k)}(x)]^+ + \left[\sum_{k'=1}^{k} \{\theta I_{k'} b(x) + c_{k'}^-(x)\} \cdot a^{(k-k')}(x) \right]^+ . \tag{1.5}$$

The relations (1.4) and (1.5) generate $c_k^-(x)$ recursively from $f_0(x)$. We are now in a position to see that a finite basis for the representation of the process is available for a certain class of increment distributions.

Let $a(x) = a^-(x) + a^+(x)$, and suppose that $a^-(x)$ has a decomposition

$$a^-(x-x') = \sum_1^{\Delta} a_m(x) b_m(x') \qquad \text{for} \quad x < 0,\ x' > 0, \tag{1.6}$$

with the following specifications: (A) The set of functions $a_m(x)$ are linearly independent and have negative support; (B) the functions $b_m(x)$ have positive support, are bounded, and are integrable $(0,\infty)$.

For any integrable function $k_+(x)$ with positive support, $[k_+(x) \cdot a^-(x)]^-$ is then a linear combination of the functions $a_m(x)$. The set of functions $\{a_m(x)\colon m = 1,2,\ldots \Delta\}$ provides a finite basis for the representation of the process. For from (1.4) and assumption B, every function $c_k^-(x)$ belongs to the linear vector space associated with this basis and may be written in the form

$$c_k^-(x) = \sum_{n=1}^{\Delta} c_n^{(k)} a_n(x) . \tag{1.7}$$

The function $c_k^-(x)$ has the vector representation

$$\mathbf{c}^{(k)} = (c_1^{(k)}, c_2^{(k)}, \ldots, \mathbf{c}_\Delta^{(k)}) \tag{1.8}$$

in this space. When the representation (1.7) is employed in (1.4) and (1.5) it is apparent that a set of equations connecting the vectors $\mathbf{c}^{(k)}$

results. These equations will be exhibited in the next section.

Let $a(x) = q\,a_{-}(x) + p\,a_{+}(x)$ so that $a^{-}(x) = q\,a_{-}(x)$. An example of an increment density $a_{-}(x)$ having the decomposition (1.6) is

$$a_{-}(x) = \left\{ \sum_{1}^{\Delta} \gamma_j\, \eta_j\, e^{\eta_j x} \right\} U(-x) \tag{1.9}$$

where $\gamma_j \geqslant 0$ and $\sum_{1}^{\Delta} \gamma_j = 1$. A second example is provided by the Erlang density

$$a_{-}(x) = \{(-\eta x)^{\Delta-1}\, \eta\, e^{\eta x} / (\Delta-1)!\}\, U(-x). \tag{1.10}$$

From the Binomial Theorem and (1.6) it is clear that a basis is provided by the set $\{a_n(x): n = 1,2,\ldots,\Delta\}$ where $a_n(x) = \{(-\eta x)^{n-1}\, e^{\eta x} / (n-1)!\}\, U(-x)$. More generally when $a_{-}(x)$ has the form

$$a_{-}(x) = \sum_{1}^{K} e^{\eta_j x}\, [P_{1j}(x) \cos \mu_j x + P_{2j}(x) \sin \mu_j x]\, U(-x) \tag{1.11}$$

where $P_{1j}(x)$ and $P_{2j}(x)$ are polynomials, it is easy to see that a finite set of functions $\{a_m(x)\}$ of the form $x^k \exp\{(\eta_j \pm i\mu_j)x\}$ will be available.

It must be emphasized that for this discrete-time process there is never any question of uniqueness or convergence. The equations (1.4) and (1.5) generate $c_k^{-}(x)$ recursively, summations are finite, and the convergence of all integrals is absolute.

A similar type of representation is available when the density $a_{+}(x)$ of the positive increments has a decomposition similar to (1.6). Processes on a finite interval also have a representation of finite dimensionality when either $a_{-}(x)$ or $a_{+}(x)$ has a finite decomposition. These cases will be discussed subsequently.

As usual the processes in continuous time are generated from the corresponding processes in discrete time as in Section V.3.

VII.2 Vector and matrix elements employed in the representation

The elements appearing in the set of vector equations (1.7) will now be written down. Equations (1.4) and (1.5) may be combined to eliminate $f_k(x)$, giving the single set of recursion equations

$$c_{k+1}^{-}(x) = -\left[[f_0 \cdot a^{(k)}]^{+} \cdot a^{-} \right]^{-} - \sum_{k'=1}^{k} \left[[\{c_{k'}^{-}(x) + \theta I_{k'} b(x)\} \cdot a^{(k-k')}]^{+} \cdot a^{-} \right]^{-}. \tag{2.1}$$

For any function $k_+(x)$ with positive support we have from (1.6) and assumption B of Section 1, $[k_+(x)\cdot a^-(x)]^- = \sum_1^\Delta a_n(x)\{\int_0^\infty k_+(x') \times b_n(x')dx'\}$, i.e.,

$$[k_+(x)\cdot a^-(x)]_n^- = \int_0^\infty k_+(x')b_n(x')dx'. \tag{2.2}$$

For the linear independence of the basis functions, (1.7), and (2.1), we obtain

$$c_n^{(k+1)} + \sum_{k'=1}^k \sum_{m=1}^\Delta c_m^{(k')} g_{mn}(k-k') = -g_{on}(k) - \theta \sum_{k'=1}^k I_{k'} g_{bn}(k-k'),$$

$$\tag{2.3}$$

where the matrix and vector elements are given by

$$g_{mn}(k) = \int_0^\infty dx'\, b_n(x') \int_{-\infty}^0 a^{(k)}(x'-x'')a_m(x'')dx''; \quad m,n = 1,2,\dots,\Delta,$$

$$\tag{2.4}$$

$$g_{on}(k) = \int_0^\infty dx'\, b_n(x') \int_0^\infty a^{(k)}(x'-x'')f_0(x'')dx'', \tag{2.5}$$

and

$$g_{bn}(k) = \int_0^\infty dx'\, b_n(x') \int_0^\infty a^{(k)}(x'-x'')b(x'')dx''. \tag{2.6}$$

When $a^-(x)$ has the form (1.6) one may in general normalize the functions $a_n(x)$ by requiring that $\int_{-\infty}^0 a_n^-(x)dx = 1$. From (1.4') and (1.7) one would then have that $I_k = -\Sigma c_n^{(k)}$. For $k \geqslant 1$, (2.3) becomes

$$c_n^{(k+1)} + \sum_{k'=1}^k \sum_m c_m^{(k')} g_{mn}(k-k') \tag{2.7}$$

$$= -g_{on}(k) + \theta \sum_{k'=1}^k \left\{\sum_m c_m^{(k')}\right\} g_{bn}(k-k').$$

The behaviour of the integrals defining $g_{mn}(k)$, $g_{on}(k)$, and $g_{bn}(k)$ for $a^-(x)$ of form (1.9) may be examined as follows. The integrals have the form

$$H_k = \int_0^\infty dx'\, B(x') \int_{-\infty}^{\infty} a^{(k)}(x'-x'')\, Q(x'')\, dx''$$

where $|Q(x)|$ is integrable and $|B(x)|$ has a maximum on $(0,\infty)$ since $B(x)$ in (1.11) is of the form $x^K \exp\{(-\eta \pm i\mu)x\}$. If we denote $\int |Q(x)|\,dx$ by Q_0 and $\max\{|B(x)|\}$ by B_0, we see that $H_k \leqslant B_0 Q_0$, i.e., the defining integrals are absolutely convergent. Since, moreover, for any density $a(x)$, $a^{(k)}(x-x')$ goes to zero for x and x' fixed as $k \to \infty$, it is easy to see that the elements $g_{mn}(k)$, $g_{on}(k)$, and $g_{bn}(k)$ all go to zero as $k \to \infty$. The vectors $\mathbf{c}^{(k)}$ are generated recursively by (2.7) from the vector $\mathbf{c}^{(1)}$ obtained from $c_1^-(x) = -[f_0 \cdot a^-]^-$.

The double integrals (2.4), (2.5) and (2.6) may be re-expressed as single contour integrals in the complex plane by the Parseval formula $\int_{-\infty}^{\infty} f(x)\, g(x)\, dx = (2\pi)^{-1} \int_{-\infty}^{\infty} \phi(z)\, \gamma(-z)\, dz$. When $f(x)$ and $g(x)$ are real, this becomes

$$\int_{-\infty}^{\infty} f(x)\, g(x)\, dx = \frac{1}{2\pi} \int_{-\infty}^{\infty} \phi(U)\, \bar{\gamma}(U)\, dU .$$

We obtain thereby, from (2.4), (2.5) and (2.6),

$$g_{mn}(k) = \frac{1}{2\pi} \int_{-\infty}^{\infty} dU\, \overline{\beta_n(U)}\, \alpha^k(U)\, \alpha_m(U) \tag{2.8}$$

$$g_{on}(k) = \frac{1}{2\pi} \int_{-\infty}^{\infty} dU\, \overline{\beta_n(U)}\, \alpha^k(U)\, \phi_0(U) \tag{2.9}$$

and

$$g_{bn}(k) = \frac{1}{2\pi} \int_{-\infty}^{\infty} dU\, \overline{\beta_n(U)}\, \alpha^k(U)\, \beta(U) . \tag{2.10}$$

The elements above may often be evaluated by the method of residues from these integrals.

VII.3 The ergodic distribution

The process X_k is ergodic when $\theta = 1$ and $\mu = \int_{-\infty}^{\infty} x\, a(x)\, dx$ is negative. In Chapter V we saw that when X_k is ergodic, its compensation function $C_k(x)$ approaches a limiting form $C_\infty(x)$ as $k \to \infty$. The density $c_k(x)$ and its negative component $c_k^-(x)$ will have corresponding limits, and the vector representation $\mathbf{c}^{(k)}$ obtained from (1.7) will have a limit $\mathbf{c}^{(\infty)}$. From reasoning identical with that of Section V.11 we infer that (2.7) takes the limiting form

$$c_n^{(\infty)} + \sum_{m=1}^{\Delta} c_m^{(\infty)} h_{mn} = \left\{ \sum_{m=1}^{\Delta} c_m^{(\infty)} \right\} h_{bn} \tag{3.1}$$

where $h_{mn} = \sum_1^\infty g_{mn}(k)$, and $h_{bn} = \sum_1^\infty g_{bn}(k)$. From (2.4) and (2.5), we then have

$$h_{mn} = \int_0^\infty dx'\, b_n(x') \int_{-\infty}^0 h(x'-x'')\, a_m(x'')\, dx'', \tag{3.2}$$

and

$$h_{bn} = \int_0^\infty dx'\, b_n(x') \int_0^\infty h(x'-x'')\, b(x'')\, dx''. \tag{3.3}$$

The function $h(x) = \sum_1^\infty a^{(k)}(x)$ is the extended renewal density of Chapter V. When $a(x)$ satisfies the conditions of Theorem V.6.1, the extended renewal density is bounded. If, moreover, $a_-(x)$ has the form (1.11), the functions $|a_m(x)|$ and $|b_m(x)|$ appearing in (1.6) are integrable, and the convergence of (3.2) and (3.3) follows.

The system of equations (3.1) consists of Δ equations in the Δ unknowns $\{c_m^{(\infty)} : m = 1,2,\ldots,\Delta\}$. This system of equations is homogeneous, and an additional equation is required to fix $\{c_m^{(\infty)}\}$. This is provided by the requirement that $\int_0^\infty f_\infty(x)\,dx = 1$. From $f_\infty(x) = \int c_\infty(x')h(x-x')\,dx' + c_\infty(x)$ given in (V.5.3), and (1.3'), we have

$$\int_0^\infty dx \left[\sum_1^\Delta c_n^{(\infty)} \int_{-\infty}^0 a_n(x')h(x-x')\,dx' + l_\infty \int_0^\infty b(x')h(x-x')\,dx' \right] + l_\infty = 1\,, \tag{3.4}$$

where $l_\infty = -\sum_1^\Delta c_n^{(\infty)}$.

A demonstration that (3.1) and (3.4) determine the unknown coefficients uniquely is required. We will provide this demonstration for the broad subclass of processes X_k for which the functions $b_n(x')$ in the decomposition (1.6) of $a^-(x-x')$ are positive. It may be seen that the Erlang density (1.10) has such a decomposition and that any linear combination of such Erlang densities with positive coefficients also has. Such processes have a semi-Markov or what may be called a semi-renewal structure, as is evident from the following discussion.

Consider the homogeneous process X_k^* governed by the same increment density $a(x)$. Suppose that a process sample $x_k^*(\omega)$ has a value $y > 0$ and that a transition occurs. The probability of a transition to the interval $(-\infty,0)$ is $A(-y)$, and the probability density on $(-\infty,0)$ conditioned on such a transition is $a^-(x-y)/A(-y) = \sum_1^\Delta a_m(x)\{b_m(y)/A(-y)\}$. The densities $a_m(x)$ may be regarded as corresponding to a set of artificial states $\mathfrak{I}_m$, indexed by $m=1,2,\ldots,\Delta$ and occupied only when a positive-to-negative transition has taken place. If a sample is at $y > 0$, the probability of remaining on $(0,\infty)$ is $1-A(-y)$, and the probability of going to $\mathfrak{I}_j$ is $b_j(y)$. If the mean increment $\mu = \int x\,a(x)\,dx$ is not zero, the states $\mathfrak{I}_j$ are not recurrent, for there is a finite probability that a sample at $\mathfrak{I}_j$ will be lost at infinity and never have another positive-to-negative traversal. The following theorem† is basic to the discussion.

Theorem VII.3.1. Let $\mathfrak{I} = \{\mathfrak{I}_m: m = 1,2,\ldots,\Delta\}$ be a prescribed set of states. Let $s_{mn}(k)$ for $m,n = 1,2,\ldots,\Delta$, $k \geqslant 1$, be the joint probability that a process X_k in the state $\mathfrak{I}_m$ at $k = 0$ revisits the set $\mathfrak{I}$ for the first time at the kth transition, and that the state visited is $\mathfrak{I}_n$. Let $h_{mn}(k)$ be the renewal probability that the set $\mathfrak{I}$ is visited at the kth transition at $\mathfrak{I}_n$, irrespective of other previous visits. Then (A) $\mathbf{h}(k) = \sum_{r=1}^{\infty} \mathbf{s}^{(r)}(k)$, where $\mathbf{s}^{(r)}(k)$ is the r-fold convolution in k defined by $\mathbf{s}^{(r+1)}(k) = \sum_{k'=1}^{k} \mathbf{s}^{(r)}(k')\,\mathbf{s}(k-k')$; (B) $[\mathbf{I}+\mathbf{X}(w)] = [\mathbf{I}+\sum_1^\infty \mathbf{h}(k)w^k]$ is non-singular, if $\sum_{k=1}^{\infty} \mathbf{s}(k)$ is strictly substochastic, for all $0 \leqslant |w| \leqslant 1$.

Statement A follows from the customary probabilistic argument relating the recurrence-time probabilities $s_{mn}(k)$ and the renewal probabilities. Let $\boldsymbol{\sigma}(w) = \sum_1^\infty \mathbf{s}(k)w^k$. Then the generating function of $\mathbf{s}^{(r)}(k)$ is $\boldsymbol{\sigma}^r(w)$, and $\mathbf{I} + \mathbf{X}(w) = \sum_0^\infty \boldsymbol{\sigma}^r(w)$. The matrix $\boldsymbol{\sigma}(1) = \sum_1^\infty \mathbf{s}(k)$ is the matrix describing all first returns. If there is a finite probability that the set $\mathfrak{I}$ will never be revisited, $\boldsymbol{\sigma}(1)$ is strictly substochastic and has all eigenvalues less than 1 in magnitude. Hence $\mathbf{I} - \boldsymbol{\sigma}(1)$ can have no zero eigenvalues and $[\mathbf{I}+\mathbf{X}(1)] = [\mathbf{I} - \boldsymbol{\sigma}(1)]^{-1} = \sum_0^\infty \boldsymbol{\sigma}^r(k)$ is

† A similar result may be found in Keilson and Wishart (1964b).

non-singular. Similarly, the maximum eigenvalue of $\boldsymbol{\sigma}(w)$ is less than that of $\sum_1^\infty |w|^k \mathbf{s}(k)$ which is less than 1 when $|w| \leqslant 1$, and $[\mathbf{I} + \boldsymbol{\chi}(w)] = [\mathbf{I} - \boldsymbol{\sigma}(w)]^{-1}$ is non-singular.

The discussion of the character of (3.1) may now be completed. We first note that $g_{mn}(k)$ of (2.4) is the probability that, if the homogeneous process X_k^* is in state $\mathfrak{S}_m$ at $k = 0$, X_k^* will have a positive-to-negative traversal at the $(k+1)$th transition. For $g_{mn}(k) = \left[[a_m(x) \cdot a^{(k)}(x)]^+ \cdot a^-]\right]_n$ and this interpretation follows. The element $g_{mn}(k)$ is therefore a renewal probability $h_{mn}(k+1)$ of the type discussed in Theorem 3.1. The matrix element h_{mn} appearing in (3.1) may be written as $h_{mn} = \sum_1^\infty g_{mn}(k) = \sum_1^\infty h_{mn}(k+1) = \sum_1^\infty h_{mn}(k) = \chi_{mn}(1)$, since $h_{mn}(1)$ is clearly zero. Thus (3.1) has the form $\mathbf{c}^{(\infty)}[\mathbf{I} + \boldsymbol{\chi}(1)] = -I_\infty \mathbf{h}_b$ where $I_\infty = -\sum c_m^{(\infty)}$. Because the process X_k is ergodic, the homogeneous process X_k^* has losses at infinity and by Theorem 3.1, $[\mathbf{I} + \boldsymbol{\chi}(1)]$ is non-singular. Hence

$$\mathbf{c}^{(\infty)} = -I_\infty \mathbf{h}_b[\mathbf{I} + \mathbf{h}]^{-1} \tag{3.5}$$

and the determination of I_∞ via (3.4) fixes all $c_m^{(\infty)}$.

The labour required for the numerical solution of (3.1) and (3.4) is substantial. In a subsequent section the lattice processes will be discussed, and similar results exhibited. The numerical calculation required for the lattice process is better suited to modern computing techniques.

VII.4 Passage processes on the half-line

For the processes studied in the previous section having a semi-Markov character, an explicit expansion may be exhibited for the passage-time probabilities. Let X_k^* be the homogeneous process governed by increment density $a(x)$ and initial density $f_0(x)$ and let $s^{(k)}$ be the probability that X_k^* attains a negative value for the first time at the kth transition. The passage process corresponds to the case $\theta = 0$ in (1.1). The probability $s^{(k)}$ is given for $k \geqslant 1$ by

$$s^{(k)} = \int_{-\infty}^0 dx \int_0^\infty f_{k-1}(x') a^-(x-x')\, dx'. \tag{4.0}$$

Hence, from (1.4) and (1.7)

$$s^{(k)} = -\int_{-\infty}^{0} c_k^-(x)\,dx = -\sum_n c_n^{(k)} = -\mathbf{c}^{(k)}\cdot\mathbf{1} \qquad (4.1)$$

where $\mathbf{1}$ is the vector (1,1,...,1).

The coefficients $c_n^{(k)}$ are obtained from the set of equations (2.7) with $\theta = 0$. If we replace $-c_m^{(k)}$ by $s_m^{(k)}$, $g_{mn}(k)$ by $h_{mn}(k+1)$ and $g_{on}(k)$ by $h_{on}(k+1)$ and note that $g_{mn}(0) = h_{mn}(1) = 0$, (2.7) takes the matrix form

$$\mathbf{s}^{(k)} + \sum_{1}^{k} \mathbf{s}^{(k')}\mathbf{h}(k-k') = \mathbf{h}_0(k), \qquad \text{for } k \geqslant 1. \qquad (4.2)$$

The semi-renewal character of (4.2) is clear. With the renewal probability interpretation of $\mathbf{h}$ and $\mathbf{h}_0$ given in the previous section, (4.2) states that for the homogeneous process X_k^*, renewals at state $\mathfrak{I}_n$ from $f_0(x)$ at time k consist of the first arrivals to $\mathfrak{I}_n$ at time k and the renewals at $\mathfrak{I}_n$ subsequent to these first arrivals.

Equation (4.2) is a vector analogue in discrete time of the equation

$$h_0(\tau) = s_0(\tau) + \int_0^{\tau} s_0(\tau')h(\tau-\tau')\,d\tau' \qquad (4.2')$$

describing a *general renewal process* in one dimension, for which the time of occurrence of the first event has a density $s_0(\tau)$ which may differ from the recurrence-time density $s(\tau)$† (see e.g. Smith, 1958). In this equation $h(\tau) = \sum_{r=1}^{\infty} s^{(r)}(\tau)$ is the ordinary renewal density and $h_0(\tau)$ is the renewal density for the general renewal process. Neither $\int s_0(\tau)d\tau = 1$, nor $\int s(\tau)d\tau = 1$ are required, i.e., there need not be a first event nor need an event recur. The unique solution to this equation is the Neumann expansion $s_0(\tau) = h_0(\tau) + \sum (-1)^r h_0(\tau) * h^{(r)}(\tau)$ which always converges (cf. Theorem V.6.4). Similarly the unique solution of (4.2) is the Neumann expansion

$$\mathbf{s}^{(k)} = \mathbf{h}_0(k) + \sum_{1}^{\infty} (-1)^r \{\mathbf{h}_0(k) * \mathbf{h}^{(r)}(k)\} \qquad (4.3)$$

where $\mathbf{a}^{(k)} * \mathbf{b}^{(k)}$ denotes the convolution $\sum_{0}^{k} \mathbf{a}^{(k')}\mathbf{b}^{(k-k')}$ and $\mathbf{h}^{(r)}(k)$ is is the r-fold convolution in k. The convergence of this equation is

† The vector $\mathbf{s}^{(k)}$ in (4.2) is the analogue of $s_0(\tau)$ in (4.2'). The recurrence-time density for the vector process would be a matrix.

assured by the Volterra character. By this we mean that $\mathbf{h}^{(1)}(k)$ has only positive support, and in particular, no support at $k = 0$. As a consequence, $\mathbf{h}^{(r)}(k) = 0$ for $0 \leqslant k \leqslant r$ and the series $\sum_{r=1}^{\infty} \mathbf{h}^{(r)}(k)$ terminates for any fixed value of k. (For the present case $\mathbf{h}^{(1)}(1)$ is also zero and $\mathbf{h}^{(r)}(k) = 0$ for $0 \leqslant k < 2r$.)

The probabilities $s^{(k)} = \mathbf{s}^{(k)} \cdot \mathbf{1}$ may be computed directly from (4.3) and the matrix elements (2.4), (2.5) and (2.6). It may again be remarked that the lattice analogue of (4.3) is more tractable numerically.

In one simple and correspondingly interesting case a closed expression is available for (4.3), provided by the following theorem.

Theorem VII.4.1. Let the sequence of independent increments ξ_i have a common absolutely continuous distribution with density

$$a(x) = q\eta\, e^{\eta x}\, U(-x) + p\, a_+(x).$$

Let $s_y^{(k)}$ be the probability that the homogeneous process $X_k = y + \sum_1^k \xi_i$, where $y > 0$, reaches $(-\infty, 0)$ for the first time at the kth transition. Then for $k \geqslant 1$,

$$s_y^{(k)} = \left(1 + \eta^{-1} \frac{d}{dy}\right)\left\{\frac{y}{k}\, a^{(k)}(-y)\right\}. \tag{4.4}$$

For the associated process in continuous time (Section II.2) where increments occur at Poisson epochs of frequency ν, the passage-time density for a first visit to $(-\infty,0)$ from $y > 0$ is (Keilson, 1964b):

$$s_y(\tau) = \left(1 + \eta^{-1} \frac{d}{dy}\right)\left\{\frac{y}{\tau}\, g(-y,\tau)\right\} \tag{4.5}$$

where $g(x,t)$ is the Green density $\delta(x)e^{-\nu t} + \sum_1^{\infty}\{e^{-\nu t}(\nu t)^k/k!\}\, a^{(k)}(x)$.

The proof of (4.4) is very similar to that of (IV.5.10) of which the result (4.4) is an extension. For the exponential density, $\Delta = 1$ and (4.2) has only one component. The generating function of (4.2) is

$$\{1 + w\,\gamma_{11}(w)\}\,\sigma(w) = w\,\gamma_{01}(w) \tag{4.6}$$

where, from (2.8), $\alpha_1(z) = \eta(\eta + iz)^{-1}$ and $\overline{\beta}_1(z) = q(\eta + iz)^{-1}$,

$$\gamma_{11}(w) = \frac{1}{2\pi}\int_{-\infty}^{+\infty} \frac{q\eta(\eta + iz)^{-2}}{1 - w\,\alpha(z)}\, dz\,; \tag{4.7}$$

and from (2.9)

$$\gamma_{01}(w) = \frac{1}{2\pi}\int_{-\infty}^{\infty} \frac{q(\eta + iz)^{-1}\phi_0(z)}{1 - w\alpha(z)}\, dz\,. \tag{4.8}$$

The integrals in (4.8) will converge if $\phi_0(z)$ goes to zero as $z \to \pm\infty$. The function $1 - w\alpha(z) = 1 - w\{q\eta(\eta + iz)^{-1} + p\alpha_+(z)\}$ has one simple pole in the half-plane $V = \mathrm{Im}(z) > 0$ and hence by Rouché's theorem (since $\alpha_+(z)$ goes to zero at infinity in the upper half-plane) has one simple zero there when $|w| < 1$. We denote this zero by $\psi(w)$. To evaluate (4.7) and (4.8) we close the integrals in the upper half-plane and find by the method of residues ($\phi_0(z)$ is analytic in the upper half-plane) that

$$1 + w\gamma_{11}(w) = i\zeta(w)q\eta\{\eta + i\psi(w)\}^{-2} \tag{4.9}$$

where $\zeta(w) = -\{\alpha'(z)\}^{-1}_{z=\psi(w)}$, and

$$w\gamma_{01}(w) = i\zeta(w)\phi_0(\psi(w))q\{\eta + i\psi(w)\}^{-1}. \tag{4.10}$$

Hence from (4.6), (4.9) and (4.10), we find

$$\sigma(w) = \phi_0(\psi(w))\{1 + i\eta^{-1}\psi(w)\}. \tag{4.11}$$

We may now permit the initial distribution $F_0(x)$ to become localized about y, so that $\phi_0(z)$ becomes e^{izy}. It follows from (4.11) that $\sigma(w) = (1 + \eta^{-1}d/dy)e^{iy\psi(w)}$, and

$$s_y^{(k)} = \left(1 + \eta^{-1}\frac{d}{dy}\right)\frac{1}{2\pi i}\oint w^{-(k+1)}\exp\{iy\psi(w)\}\,dw \tag{4.12}$$

for any contour inside the unit circle going about the origin in the positive sense. We next compare (4.12) with $a^{(k)}(-y)$. The generating function is

$$\sum_1^{\infty} w^k a^{(k)}(-y) = \frac{w}{2\pi}\int_{-\infty}^{\infty} \frac{e^{izy}\alpha(z)}{1 - w\alpha(z)}\,dz = i\zeta(w)e^{iy\psi(w)}w^{-1}$$

so that

$$y\,a^{(k)}(-y) = \frac{1}{2\pi i}\oint \exp\{iy\psi(w)\}\,iy\,\zeta(w)\,w^{-(k+2)}\,dw\,. \tag{4.13}$$

Since, however, $1 - w\alpha(\psi(w)) = 0$, differentiation with respect to w gives $\zeta(w) = w^2\psi'(w)$. An integration of

$$(2\pi i)^{-1}\oint w^{-k}(d/dw)\exp\{iy\psi(w)\}\,dw$$

by parts gives

$$(2\pi i)^{-1}\oint w^{-k-2}iy\,\zeta(w)\exp\{iy\,\psi(w)\}dw = (2\pi i)^{-1}\oint kw^{-k-1}\exp\{iy\,\psi(w)\}dw.$$

A comparison of (4.12) and (4.13) then gives the identification (4.4).

The second part of the theorem is an immediate consequence of $s_y(t) = \sum_1^\infty\{\nu(\nu t)^{k-1}\,e^{-\nu t}/(k-1)!\}s_y^{(k)}$ and $g(-y,t) = \sum_0^\infty\{e^{-\nu t}(\nu t)^k/k!\}\times g^{(k)}(-y)$.

VII.5 The modified procedure required when the positive increments have exponential form

When the component $a^+(x)$ of the increment density $a(x)$ has the form (1.11) and decomposes as in (1.6) a representation of finite dimension is again available for a replacement process X_k on $[0,\infty)$. The procedure, however, is quite distinct from that of Section 1.

As in that section the density of X_k on $[0,\infty)$ is determined by the initial density $f_0(x)$ and the continuity equations

$$f_k(x) = \int_0^\infty f_{k-1}(x')a(x',x)\,dx' \tag{5.1}$$

where $a(x',x) = a(x-x')U(x) + \theta A(-x')b(x)$. Instead of working with the sequence of equations (5.1) it is convenient to deal instead with the equations

$$f_k^*(x) = \int_{-\infty}^\infty f_{k-1}^*(x')a^*(x',x)\,dx' \tag{5.2}$$

where

$$a^*(x',x) = a(x-x')\{1-U(-x')U(x)\} + \theta A(-x')b(x)U(x'). \tag{5.3}$$

The functions $f_k^*(x)$ have support on the entire real line. The transition density $a^*(x',x)$ need not be a probability density even when $\theta = 1$, for when $x' > 0$, $\int_{-\infty}^\infty a^*(x',x)\,dx = 1 + \theta A(-x') \geqslant 1$. Consequently $f_k^*(x)$ does not have a valid probabilistic meaning for all x. Its positive component $f_k^*(x)U(x)$, however, does, for, as we will now show, when $f_0^*(x) = f_0(x)$,

$$f_k^*(x)U(x) = f_k(x), \qquad \text{for all } k. \tag{5.4}$$

The demonstration is by induction. Clearly $a^*(x',x) = a(x',x)$ for x' and x both positive. Moreover $a^*(x',x) = 0$ when $x' < 0$ and simultaneously $x > 0$. Hence the validity of (5.4) for any k implies its validity for $k+1$.

The sequence of functions $f_k^*(x)$ generated by (5.2) are readily seen to be positive and integrable. For an ergodic process, moreover, the sequence of functions $f_k^*(x)$ will be shown subsequently to have a finite limit as $k \to \infty$ for all x.

From (5.2) and (5.3) we may write

$$f_k^*(x) = f_{k-1}^*(x) \cdot a(x) + \theta I_k\, b(x) + c_k(x), \tag{5.5}$$

where

$$c_{k+1}(x) = \left\{ -\int_{-\infty}^{0} f_k^*(x')\,a(x-x')\,dx' \right\} U(x) \tag{5.6}$$

and

$$I_{k+1} = \int_0^\infty f_k(x')\,A(-x')\,dx'. \tag{5.7}$$

From (5.5) one obtains

$$f_k^*(x) = f_0(x) \cdot a^{(k)}(x) + \theta \sum_1^k I_{k'} \{b(x) \cdot a^{(k-k')}(x)\}$$

$$+ \sum_1^k c_{k'}(x) \cdot a^{(k-k')}(x) \tag{5.8}$$

and this may be substituted into (5.6). There results

$$c_{k+1}(x) + \sum_{k'=1}^{k} \left[[c_{k'}(x) \cdot a^{(k-k')}(x)]^- \cdot a(x) \right]^+ = \tag{5.9}$$

$$= -\left[[f_0(x)\cdot a^{(k)}(x)]^{-}\cdot a(x)\right]^{+} - \theta\sum_{k'=1}^{k}\left[[I_{k'}\,b(x)\cdot a^{(k-k')}(x)]^{-}\cdot a\right]^{+}$$

and a similar equation for I_k obtained from (5.7) and (5.8). For discussion of the ergodic distribution or a passage process ($\theta = 0$), this equation is not needed.

It may now be observed from (5.6) that a finite basis for the functions $c_k(x)$ will be available when $a^+(x)$ has the decomposition (cf. (1.6)):

$$a^+(x-x') = \sum_{m=1}^{\Delta} a_m(x)\,b_m(x'), \qquad x' < 0,\ x > 0 \qquad (5.10)$$

with the specification that (A) the functions $a_m(x)$ are linearly independent and have positive support, and (B) the functions $b_m(x)$ have negative support and are absolutely integrable on $(-\infty,0)$.

The set of functions $\{a_m(x)\colon m = 1,2,\ldots,\Delta\}$ form a basis of finite dimension for the representation of (5.9). In this basis we may write $c_k(x) = \sum_1^{\Delta} c_m^{(k)}\,a_m(x)$ and represent $c_k(x)$ by $\mathbf{c}^{(k)} = \{c_1^{(k)}, c_2^{(k)}, \ldots, c_\Delta^{(k)}\}$. With this representation, (5.9) becomes

$$c_n^{(k+1)} + \sum_{k'=1}^{k}\sum_{m} c_m^{(k')}\,g_{mn}(k-k')$$

$$= -g_{on}(k) - \theta\sum_{k'=1}^{k} I_{k'}\,g_{bn}(k-k') \qquad (5.11)$$

where

$$g_{mn}(k) = \int_{-\infty}^{0} dx'\,b_n(x')\int_0^{\infty} a^{(k)}(x'-x'')\,a_m(x'')\,dx'', \qquad (5.12)$$

$$g_{on}(k) = \int_{-\infty}^{0} dx'\,b_n(x')\int_0^{\infty} a^{(k)}(x'-x'')\,f_0(x'')\,dx'', \qquad (5.13)$$

and

$$g_{bn}(k) = \int_{-\infty}^{0} dx'\,b_n(x')\int_0^{\infty} a^{(k)}(x'-x'')\,b(x'')\,dx''. \qquad (5.14)$$

We will now show that, if the process is ergodic, $\mathbf{c}^{(k)}$ approaches a limiting vector $\mathbf{c}^{(\infty)}$ as $k \to \infty$. From (5.11) with $\theta = 1$, we have for the generating function $\mathbf{c}(w) = \sum_0^\infty \mathbf{c}^{(k)} w^k$

$$w^{-1}\,\mathbf{c}(w)\{\mathbf{I} + w\,\boldsymbol{\gamma}(w)\} = -\boldsymbol{\gamma}_0(w) - i(w)\,\boldsymbol{\gamma}_b(w) \tag{5.15}$$

and

$$w^{-1}\,\mathbf{c}(w) = -\boldsymbol{\gamma}_0(w)\{\mathbf{I} + w\,\boldsymbol{\gamma}(w)\}^{-1} - i(w)\,\boldsymbol{\gamma}_b(w)\{\mathbf{I} + w\,\boldsymbol{\gamma}(w)\}^{-1} \tag{5.16}$$

where $\boldsymbol{\gamma}(w)$ and $\boldsymbol{\gamma}_0(w)$ are the generating functions obtained from (5.12) and (5.13). Equation (5.16) may now be multiplied by $(1-w)$ and w passed to 1 along an arbitrary path inside the unit circle. We saw from Theorem 3.1 that when $\int x a(x)\,dx \neq 0$, the matrix $\{\mathbf{I} + \boldsymbol{\chi}(w)\} = \{\mathbf{I} + w\,\boldsymbol{\gamma}(w)\}$ governing (4.2) is non-singular at $w = 1$. The identical reasoning may be applied to the negative-to-positive traversals associated with the decomposition (5.10), to infer that $\{\mathbf{I} + w\,\boldsymbol{\gamma}(w)\}$ has an inverse for $w = 1$. We also have for (5.16) $\lim\,(1-w)\,\boldsymbol{\gamma}_0(w) = 0$, since $g_{on}(k) \to 0$ as $k \to \infty$, and $\lim\,(1-w)\,i(w) = I_\infty$ from the assumption of ergodicity and (5.7). Hence from (5.16) we have $\mathbf{c}^{(\infty)} = \{-I_\infty \boldsymbol{\gamma}_b(1)\} \times \{\mathbf{I} + \boldsymbol{\gamma}(1)\}^{-1}$. Equivalently the coefficients $c_n^{(\infty)}$ may be found as the solution of the set of equations

$$c_n^{(\infty)} + \sum_1^{\Delta} c_m^{(\infty)} h_{mn} = -I_\infty\, h_{bn} \tag{5.17}$$

where

$$h_{mn} = \int_{-\infty}^{0} b_n(x') \int_0^{\infty} h(x'-x'')\, a_m(x'')\, dx''\, dx' \tag{5.18}$$

and

$$h_{bn} = \int_{-\infty}^{0} b_n(x') \int_0^{\infty} h(x'-x'')\, b(x'')\, dx''\, dx'. \tag{5.19}$$

The function $h(x)$ in (5.18) and (5.19) is the extended renewal density $h(x) = \sum_1^\infty a^{(m)}(x)$ of Chapter V. The coefficient I_∞ in (5.17) may be obtained from normalization. Thus from (5.8), since $\theta = 1$,

$$f_\infty^*(x) = \int_0^\infty h(x-x')\left\{I_\infty b(x') + \sum c_m^{(\infty)} a_m(x')\right\} dx' \qquad (5.20)$$

and since $f_\infty(x) = f_\infty^*(x)U(x)$, the normalization equation reads

$$\sum c_m^{(\infty)} \int_0^\infty \int_0^\infty h(x-x')a_m(x')\,dx\,dx'$$
$$+ I_\infty \int_0^\infty \int_0^\infty h(x-x')\,b(x')\,dx\,dx' = 1. \qquad (5.21)$$

A similar procedure permits one to treat processes on the finite interval whenever either $a^-(x)$ has the decomposition (1.6) or $a^+(x)$ has the decomposition (5.10). A discussion for processes in continuous time may be found in Keilson (1964b). The method is quite similar to that employed in Section 7 to discuss lattice processes on a finite interval.

VII.6 Processes on the half-lattice, with bounded increments

The replacement processes of Section V.4 on the set of states $\{n:\ 0 \leqslant n < \infty\}$ have similar finite representations for two distinct types of increment distributions.

In the first type the increments of at least one sign are bounded. For example, the negative increments are bounded if $e_n = P\{\xi_j = n\} = 0$ for $n < -N_1$. Similarly the positive increments are bounded if $e_n = 0$ for $n > N_2$. If we permit the bounds $-N_1$ and N_2 to be infinite, a pair of numbers $-N_1$ and N_2 may always be assigned to any lattice increment distribution. The dimension of the most economical representation available is then given by the following theorem.

Theorem VII.6.1. For any passage process or replacement process on the lattice of non-negative integers whose governing increment distribution has one or both bounds finite, a finite representation is available for the discussion of the process. The dimension of the representation is $\Delta = \min[N_1, N_2]$.

We are treating the replacement process N_k on the state space $\{n:\ 0 \leqslant n < \infty\}$ augmented by an absorbing state $\mathfrak{U}$ and governed by the continuity equation

$$p_n^{(k)} = \sum p_{n'}^{(k-1)} e_{n'n}\,. \qquad (6.1)$$

In (6.1) $e_{n'n}$ is the single-step transition probability

$$\mathrm{e}_{n'n} = P\{N_k = n | N_{k-1} = n'\} = \mathrm{e}_{n-n'} U(n) + \theta E_{-n'-1}\, b_n U(n'), \qquad (6.2)$$

with $E_{-n} = \sum_{-\infty}^{-n} \mathrm{e}_j$. Equation (6.2) states that when $N_{k-1} + \xi_k < 0$, the process is replaced at n on $[0,\infty)$ with probability θb_n or terminates at $\mathfrak{U}$ with probability $(1-\theta)$. An initial value m is prescribed. When $\theta = 1$ and $\mu = \Sigma n\, \mathrm{e}_n < 0$ the process is ergodic. When $\theta = 0$, we have the passage process.

The method of treatment is similar to that of Section 5. Let $N = \min(N_1, N_2)$. Instead of working with (6.1) we introduce $p_n^{*(k)}$ for $-\infty < n < \infty$, $k \geqslant 0$, defined by $p_n^{*(0)} = p_n^{(0)}$ and

$$p_n^{*(k)} = \sum_{-\infty}^{\infty} p_{n'}^{*(k-1)} \mathrm{e}_{n'n}^{*} \qquad (6.3)$$

where

$$\mathrm{e}_{n'n}^{*} = \mathrm{e}_{n-n'}\{1 - U_A(n') U_B(n)\} + \theta E_{-n'-1} b_n U(n'). \qquad (6.4)$$

The functions $U_A(n)$ and $U_B(n)$ are set functions. $U_A(n) = 1$ when n is on the set $A = \{n\colon n < -N \text{ or } n \geqslant 0\}$ and $U_A(n) = 0$ otherwise. $U_B(n) = 1$ when n is on the set $B = \{n\colon -N \leqslant n \leqslant -1\}$ and $U_B(n) = 0$ otherwise. The first term in (6.4) permits all increments except those that would carry a sample to set B. The second term replaces samples ω for which $n_k(\omega) + \xi_{k+1}$ is negative. As in Section 5 one infers by induction that when $p_n^{*(0)} = p_n^{(0)}$

$$p_n^{*(k)} = p_n^{(k)} \qquad \text{for } n \geqslant 0,\ k \geqslant 0. \qquad (6.5)$$

To demonstrate this we will show that (I) $\mathrm{e}_{n'n}^{*} = \mathrm{e}_{n'n}$ for n', n both on $[0,\infty)$; and (II) $p_{n'}^{*(k)} \mathrm{e}_{n'n}^{*} = 0$ when $n' < 0$ and simultaneously $n \geqslant 0$. Part I is apparent from (6.2) and (6.4).

To see Part II, we note that $p_n^{*(k)} = 0$ for n in B for all k. This follows from $\mathrm{e}_{n'n}^{*} = 0$ for n' in A, n in B, (6.3) and induction on k. If $N_1 \geqslant N_2$ so that $N = N_2$ we then have for $n \geqslant 0$, $\sum_{n'<0} p_{n'}^{*(k)} \mathrm{e}_{n'n}^{*} = \sum_{n' \in B} p_{n'}^{*(k)} \mathrm{e}_{n'n}^{*} = 0$. If $N_1 \leqslant N_2$ so that $N = N_1$, $p_n^{*(k)} = 0$ for all $n < 0$. For, $p_n^{*(k)} = 0$ for n in B and hence $p_n^{*(k)} = 0$ for $n < -N_1$ from (6.3). We then have $\sum_{n'<0} p_{n'}^{*(k)} \mathrm{e}_{n'n}^{*} = 0$ and II has been verified. From (6.3), I, and II, (6.5) follows by induction.

From (6.3) and (6.4) we may write as in Section 5,

$$p_n^{*(k)} = \sum_{-\infty}^{\infty} p_{n'}^{*(k-1)} e_{n-n'} + \theta I_k b_n + c_n^{*(k)} \tag{6.6}$$

where

$$c_n^{*(k+1)} = -\left\{\sum_A p_{n'}^{*(k)} e_{n-n'}\right\} U_B(n) = -\left\{\sum_{-\infty}^{+\infty} p_{n'}^{*(k)} e_{n-n'}\right\} U_B(n) \tag{6.7}$$

and

$$I_{k+1} = \sum_0^{\infty} p_{n'}^{*(k)} E_{-n'-1} \; . \tag{6.8}$$

A set of N equations for the compensation functions $c_n^{*(k)}$, $n = -1,-2, \ldots,-N$, may now be obtained. From (6.6) we have, for all n, when $p_n^{(0)} = \delta_{n,n_0}$,

$$p_n^{*(k)} = g_{n-n_0}^{(k)} + \theta \sum_{k'=1}^{k} \sum_{n'=0}^{\infty} I_{k'} b_{n'} g_{n-n'}^{(k-k')} + \sum_{k'=1}^{k} \sum_B c_{n'}^{*(k')} g_{n-n'}^{(k-k')} . \tag{6.9}$$

We now substitute (6.9) into (6.7) and take advantage of the simplifying relation $g_n^{(k+1)} = \sum_{n'} e_{n-n'} g_{n'}^{(k)}$ valid for all k, to obtain for all n in set B

$$c_n^{*(k+1)} + \sum_B \sum_{k'=1}^{k} g_{n-n'}^{(k+1-k')} c_{n'}^{*(k')} = -g_{n-n_0'}^{(k+1)} - \theta \sum_0^{\infty} \sum_{k'=1}^{k} g_{n-n'}^{(k+1-k')} I_{k'} b_{n'}, \tag{6.10}$$

and this is the desired set of recursive equations governing $c_n^{*(k)}$. It must be supplemented by the recursion equations governing I_k obtained from (6.8) and (6.9). This equation will not be required.

Equation (6.10) serves as a jumping-off point for the treatment of the passage process ($\theta = 0$) and ergodic process ($\theta = 1$). We treat the former first. Suppose that $N_1 < N_2$. In our demonstration of (6.2) we showed that $p_n^{*(k)} = 0$ for $n < -N_1$ for this case. Consequently, from (6.7) we see that $-c_n^{*(k+1)}$ is the probability that a sample starting at n_0 will reach $(-\infty,-1]$ for the first time at the $(k+1)$th transition and that the value at that time will be n. The passage-time probability is then

$$s_{n_0}^{(k)} = -\sum_B c_{n'}^{*(k)} . \tag{6.10'}$$

For this case, (6.10) becomes

$$s_{n_0 n}^{(k+1)} + \sum_B \sum_{k'=1}^{k} g_{n-n'}^{(k+1-k')} s_{n_0 n'}^{(k')} = g_{n-n_0}^{(k+1)} \tag{6.11}$$

and this equation has the following simple renewal interpretation in terms of the behaviour of the associated homogeneous process. (For simplicity we assume $e_0 = 0$.) The renewal probability for all visits to the set B at state n at the $(k+1)$th epoch is $g_{n-n_0}^{(k+1)}$. Such visits consist of first visits with probability $s_{n_0 n}^{(k+1)}$ and visits subsequent to a previous first visit to B at n' at time k'. The Volterra character of the matrix of renewal probabilities, discussed below (4.2), again assures the convergence of the Neumann series (4.3). This series now takes the form, for $k \geqslant 1$,

$$\mathbf{s}^{(k)} = \mathbf{g}_0^{(k)} + \sum_1^{\infty} (-1)^r \mathbf{g}_0^{(k)} * \{\mathbf{g}^{(k)} \overset{*}{} \mathbf{g}^{(k)} \overset{*}{} \ldots \mathbf{g}^{(k)}\} \tag{6.12}$$

r-fold convolution in k,

where $\mathbf{s}^{(k)}$ and $\mathbf{g}_0^{(k)}$ are vectors with components $(\mathbf{s}^{(k)})_n = s_{n_0 n}^{(k)}$ and $(\mathbf{g}_0^{(k)})_n = g_{n-n_0}^{(k)}$, and $\mathbf{g}^{(k)}$ is the Toeplitz matrix with elements $(\mathbf{g}^{(k)})_{mn} = g_{n-m}$. The passage-time probabilities are thereby exhibited in terms of the Green probabilities.

The functions $c_n^{*(k)}$ give the passage-time probabilities to the set B for the homogeneous process, but when $N_1 > N_2$ this is not the same as the passage-time probabilities to the set $(-\infty, 0)$. It is clear, however, that we may employ the relation $s_{n_0}^{(k)} = \sum_{-\infty}^{-1} \sum_{n'=0}^{\infty} p_{n'}^{*(k-1)} e_{n-n'}$ with (6.9) to obtain the proper result in slightly more complicated form.

The ergodic distribution may be written down from (6.9). We have when $\theta = 1$ and $\Sigma n e_n = \mu < 0$, by the reasoning of Section V.11,

$$p_n^{*(\infty)} = I_\infty \sum_0^\infty b_{n'} g_{n-n'} + \sum_B c_{n'}^{*(\infty)} g_{n-n'} \tag{6.13}$$

where $g_n = \delta_{n,0} + h_n = \delta_{n,0} + \sum_{r=1}^{\infty} e_n^{(r)}$ is the ergodic Green mass for the lattice of Section V.7. The compensation numbers $c_{n'}^{*(\infty)}$, $n' = -1,-2,\ldots -N$, are obtained from the limiting form of equation (6.10),

$$c_n^{*(\infty)} + \sum_B c_{n'}^{*(\infty)} h_{n-n'} = -I_\infty \sum_0^\infty b_{n'} h_{n-n'} \tag{6.14}$$

and the normalizing equation $\sum_0^\infty p_n^{(\infty)} = \sum_0^\infty p_n^{*(\infty)} = 1$. From (6.13), this reads

$$I_\infty \left\{ \sum_{n=0}^\infty \sum_{n'=0}^\infty b_{n'} g_{n-n'} \right\} + \sum_B c_{n'}^{*(\infty)} \sum_{n=0}^\infty g_{n-n'} = 1. \tag{6.15}$$

These two equations determine $c_n^{*(\infty)}$ and I_∞ uniquely. For the matrix $\{I + h\}$ appearing on the left of (6.14) is non-singular by virtue of Theorem 3.1. From (6.14) $c_n^{*(\infty)}$ may be expressed uniquely as $c_n^{*(\infty)} = \kappa_n I_\infty$ and I_∞ may then be obtained from (6.15).

It should be noted that the representation used is not unique. If, for example, the increment distribution had $N_1 = 4$ and $N_2 = 2$, and the passage probability was of interest, one could choose to work with $N = 4$ instead of the smaller value 2, so that the compensation functions $c_n^{*(k)}$ would have direct meaning as passage-time probability components and (6.10′) would be valid.

A second type of structure for the set $\{e_n\}$ of increment probabilities also gives rise to a finite representation. This type has a structure which is the analogue of (1.11). If for either positive or negative increments, e_n has the form

$$e_n = \sum_{k=1}^K \beta_k^{|n|} [P_{1k}(n) \cos \mu_k n + P_{2k}(n) \sin \mu_k n] \tag{6.16}$$

where $\mu_1 = 0$ and $0 < \beta_k < 1$, a representation of finite dimension is available. The finite representation results from the structure-preserving character of (6.16). The simplest structure-preserving set $\{e_n\}$ for $K = 1$ has $e_n = \alpha\beta^n$ for $n < 0$, say, with $\sum_0^\infty e_{n-m} p_m =$

$\alpha\beta^n\{\sum_0^\infty p_m\,\beta^m\}$ for $n < 0$. The discussion required is a direct analogue of that given for the continuum in Sections 1 through 5, and will not be presented.

VII.7 Processes on the finite lattice

The discussion of the previous section may be extended immediately to processes on a finite interval for which the increments of at least one sign are bounded. Instead of the dimension $\Delta = \min\{N_1, N_2\}$ obtained for processes on the half-lattice, one finds a representation available with a dimension twice as large.

Theorem VII.7.1. For any replacement process on the finite lattice $\{n\colon 0 \leqslant n \leqslant L\}$ whose governing increment distribution has one or both bounds finite, a finite representation is available for the discussion of the process. The dimension of the representation is $\Delta = 2N$ where $N = \min\{N_1, N_2\}$.

In place of the compensation structure of Section 6, we will be able to discuss the process by a set of compensation functions $c_n^{*(k)}$ on $B = \{n\colon n = -N, -N+1, \ldots, -1,\ L+1, L+2, \ldots, L+N\}$. For we may again replace the process $p_n^{(k)} = \Sigma\, p_{n'}^{(k-1)} e_{n'n}$, where

$$e_{n'n} = e_{n-n'}\{U(n) - U(n-L-1)\} + \theta_1 E_{-n'-1}\, b_{1n} + \theta_2(1 - E_{L-n'})\, b_{2n} \tag{7.1}$$

with

$$p_n^{*(k)} = \sum p_{n'}^{*(k-1)}\, e^*_{n'n} \tag{7.2}$$

where

$$e^*_{n'n} = e_{n-n'}\{1 - U_A(n')\,U_B(n)\}$$
$$+ \{\theta_1 E_{-n'-1}\, b_{1n} + \theta_2(1 - E_{L-n'})\, b_{2n}\}\{U(n') - U(n'-L-1)\}. \tag{7.3}$$

In (7.3), $U_A(n)$ and $U_B(n)$ are set functions defined as in Section 6. B is the set given above and A is the set of all points on the lattice not in B. As in Section 6, one infers that

$$p_n^{*(k)} = p_n^{(k)} \qquad \text{for} \quad 0 < n \leqslant L,\ k \geqslant 0 \tag{7.4}$$

by the same type of reasoning employed there. We then find from (7.2) and (7.3) that

$$p_n^{*(k)} = \sum_{-\infty}^{\infty} p_{n'}^{*(k-1)} e_{n-n'} + \theta_1 I_{1k} b_{1n} + \theta_2 I_{2k} b_{2n} + c_n^{*(k)} \qquad (7.5)$$

where

$$c_n^{*(k+1)} = -\left\{ \sum_A p_{n'}^{*(k)} e_{n-n'} \right\} U_B(n) \qquad (7.6)$$

and, for $k \geqslant 1$,

$$\text{(a)}\ I_{1k} = \sum_0^L p_{n'}^{*(k-1)} E_{-n'-1}; \quad \text{(b)}\ I_{2k} = \sum_0^L p_{n'}^{*(k-1)} (1-E_{L-n'}). \qquad (7.7)$$

The analogue of (6.9) is then, for $0 \leqslant n_0 \leqslant L$,

$$p_n^{*(k)} = g_{n-n_0}^{(k)} + \theta_1 \sum_1^k \sum_0^L b_{1n'} g_{n-n'}^{(k-k')} I_{1k'}$$

$$+ \theta_2 \sum_1^k \sum_0^L b_{2n'} g_{n-n'}^{(k-k')} I_{2k'} + \sum_1^k \sum_B c_{n'}^{*(k')} g_{n-n'}^{(k-k')} \qquad (7.8)$$

Finally, in place of (6.10) one has for all n in B,

$$c_n^{*(k+1)} + \sum_B \sum_{k'=1}^k g_{n-n'}^{(k-k'+1)} c_{n'}^{*(k')}$$

$$= -g_{n-n_0}^{(k+1)} - \theta_1 \sum_0^L \sum_{k'=1}^k g_{n-n'}^{(k-k'+1)} I_{1k'} b_{1n'}$$

$$- \theta_2 \sum_0^L \sum_{k'=1}^k g_{n-n'}^{(k-k'+1)} I_{2k'} b_{2n'}. \qquad (7.9)$$

Several passage processes may be of interest. The simplest is that for $\theta_1 = \theta_2 = 0$. For this passage process, the functions $-c_n^{*(k)}$ are the probabilities $s_{n_0 n}^{*(k+1)}$ that the homogeneous process starting at n_0 will visit B for the first time at n, at the $(k+1)$th epoch. From these

probabilities the first passage probabilities for the inhomogeneous
process may be calculated as in Section 6. (If N_1 and N_2 were both
small, it might be more desirable to work with $B = \{n: -N_1 \leqslant n \leqslant -1$ or
$L+1 \leqslant n \leqslant L+N_2\}$ so that the compensation elements have a direct
interpretation as passage-time probabilities. The formalism will be
only slightly different.) The Neumann series (6.12) is again available.

One might also be interested in $\theta_1 = 0$ and $\theta_2 = 1$, describing
terminations when a negative value is reached or in $\theta_1 = 1$ and $\theta_2 = 0$.
Such passage processes have a more complex structure which we will
not discuss.

When $\theta_1 = \theta_2 = 1$, the process will be ergodic, whether or not
$\mu = 0$. The existence of the ergodic Green measure is required,
however, for our discussion, and the condition $\mu \neq 0$ must again be
imposed. From (7.9) we have for all n in B,

$$c_n^{*(\infty)} + \sum_B c_{n'}^{*(\infty)} h_{n-n'} = -I_{1\infty} \sum_0^L b_{1n'} h_{n-n'} - I_{2\infty} \sum_0^L b_{2n'} h_{n-n'}, \quad (7.10)$$

where h_n is the extended renewal mass $\sum_{r=1}^{\infty} e_n^{(r)}$. Similarly, from (7.7)
and (7.8),

$$p_n^{*(\infty)} = I_{1\infty} \sum_0^L b_{1n'} \, g_{n-n'} + I_{2\infty} \sum_0^L b_{2n'} \, g_{n-n'} + \sum_B c_{n'}^{*(\infty)} \, g_{n-n'}, \quad (7.11)$$

where $g_n = \delta_{n,0} + h_n$ is the ergodic Green mass, and

$$\text{(a) } I_{1\infty} = \sum_0^L p_{n'}^{*(\infty)} E_{-n'-1}; \quad \text{(b) } I_{2\infty} = \sum_0^L p_{n'}^{*(\infty)}(1-E_{L-n'}). \quad (7.12)$$

A supplementary equation is provided by the normalization condition

$$\sum_0^L p_n^{*(\infty)} = 1. \quad (7.13)$$

When $\mu \neq 0$, the visits of the homogeneous process to the set of
states B are not recurrent, and the matrix $(\mathbf{I} + \mathbf{h})$ appearing on the
left-hand side of (7.10) is non-singular by virtue of Theorem 3.1. Thus
$c_n^{*(\infty)}$ may be expressed in terms of $I_{1\infty}$ and $I_{2\infty}$, and $p_n^{*(\infty)}$ may also
be so expressed from (7.11). Equations (7.12) then give a pair of
homogeneous equations in the unknowns $I_{1\infty}$ and $I_{2\infty}$ which must be

equivalent. Either of these with (7.13) determines $I_{1\infty}$ and $I_{2\infty}$.

To illustrate the formalism, let us consider an arbitrary skip-free process on the lattice modified by two retaining boundaries at $n = 0$, and $n = L$. With no loss of generality we may treat the skip-free negative case with $e_{-1} \neq 0$, and $e_j = 0$ for $j < -1$. The sign of the mean increment μ ($\neq 0$) need not be specified. The set B consists of the two states $n = -1$ and $n = L+1$. We may write (7.10) for $n = -1$ as

$$g_0 c_{-1}^{*(\infty)} + g_{-L-2} c_{L+1}^{*(\infty)} = -I_{1\infty} g_{-1} - I_{2\infty} g_{-1-L} \tag{7.14}$$

and (7.11) becomes

$$p_n^{*(\infty)} = I_{1\infty} g_n + I_{2\infty} g_{n-L} + c_{-1}^{*(\infty)} g_{n+1} + c_{L+1}^{*(\infty)} g_{n-L-1}\,. \tag{7.15}$$

As for the justification of (6.5), it is readily seen that $p_n^{*(k)} = 0$ for all $n \leqslant 0$. Hence from (7.6) we see that

$$c_{-1}^{*(\infty)} = -p_0^{*(\infty)} e_{-1} = -I_{1\infty}\,. \tag{7.16}$$

From Theorems V.8.3′ and V.8.4′ we see that $g_n = |\mu|^{-1}$ for $\mu < 0$ for all $n \leqslant 0$, and $g_n = |\mu_{\rho_1}|^{-1} \rho_1^{-n}$ for $\mu > 0$ for all $n \leqslant 0$. Thus we may write for either case

$$g_n = |\mu_\theta|^{-1} \theta^{-n} \tag{7.17}$$

with $\theta = 1$ for $\mu < 0$ and $\theta = \rho_1 < 1$ for $\mu > 0$. If (7.16) is now introduced into (7.14), we obtain

$$\theta c_{L+1}^{*(\infty)} + I_{2\infty} = \theta^{-L-1}(1-\theta) I_{1\infty} \tag{7.18}$$

and (7.15) becomes

$$p_n^{*(\infty)} = I_{1\infty}\{(g_n - g_{n+1}) + |\mu_\theta|^{-1}\theta^{-n-1}(1-\theta)\}\,. \tag{7.19}$$

The normalization condition $\sum_0^L p_n^{*(\infty)} = 1$ may now be employed to give

$$I_{1\infty} = \{|\mu_\theta|^{-1}\theta^{-L-1} - g_{L+1}\}^{-1}. \tag{7.20}$$

If $\mu < 0$, then $\theta = 1$, and (7.19) and (7.20) reduce to the previous result (VI.4.4). When $\mu > 0$, the ergodic distribution takes the form exhibited in (VI.4.8).

BIBLIOGRAPHY

Bartlett, M.S. (1955) *An Introduction to Stochastic Processes*, Cambridge University Press.

Baxter, G. (1958) "An operator identity", *Pacific J. Math.*, **8**, 649-663.

Bharucha-Reid, A.T. (1960) *Elements of the Theory of Markov Processes and their Applications*, McGraw-Hill Book Co. Inc., New York.

deBruijn, N.G. (1958) *Asymptotic Methods in Analysis*, North-Holland Publishing Co.

Chung, K.L. and Fuchs, W.H.J. (1951) "On the distribution of values of sums of random variables", *Amer. Math. Soc.*, Memoir 6, 1-12.

Chung, K.L. (1960) *Markov Chains with Stationary Transition Probabilities*, Springer Verlag.

Cramér, H. (1938) "Sur une nouveau théorème-limite de la théorie des probabilités", *Actualités Scientifiques et Industrielles*, No. 736, Hermann et Cie, Paris.

Cramér, H. (1962) *Random Variables and Probability Distributions*, 2nd edn, Cambridge University Press.

Copson, E.T. (1935) *An Introduction to the Theory of Functions of a Complex Variable*, Clarendon Press, Oxford.

Daniels, H.E. (1954) "Saddlepoint approximations in statistics", *Ann. Math. Statist.*, **25**, 631-650.

Doob, J.L. (1953) *Stochastic Processes*, John Wiley & Sons Inc., New York.

Esscher, F. (1932) "On the probability function in the collective theory of risk", *Skand. Akt.*, **15**, 175-195.

Feller, W. (1941) "On the integral equation of renewal theory", *Ann. Math. Statist.*, **12**, 243-267.

Feller, W. (1957) *An Introduction to Probability Theory and its Applications*, John Wiley & Sons Inc., New York, 2nd edn.

Feller, W. (1959) *On Combinatorial Methods in Fluctuation Theory*, The Harald Cramér Volume, J. Wiley, N.Y.

Feller, W. and Orey, S. (1961) "A renewal theorem", *J. Math. and Mech.* **10**, 619-624.

Finch, P.D. (1960) "On the transient behavior of a simple queue", *J. Roy. Statist. Soc.*, **22**, 277-284.

Fisz, M. (1963) *Probability Theory and Mathematical Statistics*, John Wiley & Sons, New York.

Gaver, D.P., Jr. and R.G. Miller, Jr. (1962) "Limiting distributions for some storage problems", *Studies in Applied Probability and Management Science* (ed. by K.J. Arrow, S. Karlin, and H. Scarf), Stanford University Press, 110-126.

Gnedenko, B. V. and Kolmogorov, A.N. (1954) *Limit Distributions for Sums of Independent Random Variables*, Addison-Wesley, Cambridge, Mass.

Good, I.J. (1957) "Saddlepoint methods for the multinomial distribution", *Ann. Math. Statist.*, **28**, 861-881.

Jeffreys, H. and Jeffreys, B.S. (1962) *Methods of Mathematical Physics*, 3rd edn, Cambridge University Press.

Keilson, J. (1961) "The homogeneous random walk on the half-line and the Hilbert problem", *Bull. I.S.I.*, No. 113, 33e Session, Paris.

Keilson, J. (1962a) "The use of Green's functions in the study of bounded random walks with application to queuing theory", *J. Math. and Phys.*, **41**, 42-52.

Keilson, J. (1962b) "Non-stationary Markov walks on the lattice", *J. Math. and Phys.*, **41**, 205-211.

Keilson, J. (1962c) "The general bulk queue as a Hilbert problem", *J.R. Statist. Soc.*, Ser. B, **24**, 344-358.

Keilson, J. (1963a) "The first passage time density for homogeneous skip-free walks on the continuum", *Ann. Math. Statist.*, **34**, 1003-1011.

Keilson, J. (1963b) "On the asymptotic behaviour of queues", *J. R. Statist. Soc.*, Ser. B, **25**, 464-476.

Keilson, J. (1963c) "A gambler's ruin problem in queuing theory", *Operations Research,* **11**, 570-576.

Keilson, J. (1964a) "Some comments on single-server queuing methods and some new results", *Proc. Camb. Phil. Soc.*, **60**, 237-251.

Keilson, J. (1964b) "An alternative to Wiener-Hopf methods for the study of bounded processes", *J. of Applied Probability*, **1**, 85-120.

Keilson, J. (1964c) "On the ruin problem for the generalized random walk", *Operations Research*, **12**, 504-506.

Keilson, J. and Wishart, D.M.G. (1964) "A central limit theorem for processes defined on a finite Markov chain", *Proc. Camb. Phil. Soc.*, **60**, 547-567.

Keilson, J. and Wishart, D.M.G. (1965) "Boundary problems for additive processes defined on a finite Markov chain", *Proc. Camb. Phil. Soc.* (to appear).

Kemperman, J.H.B. (1961) *The Passage Problem for a Stationary Markov Chain*, University of Chicago Press.

Kemperman, J.H.B. (1964) "A Wiener-Hopf type method for a general random walk with a two-sided boundary", *Ann. Math. Statist.*, **34**, 1168-1193.

Kendall, D.G. (1951) "Some problems in the theory of queues", *J. Roy. Statist. Soc.*, Ser. B, **13**, 151-185.

Kendall, D.G. (1953) "Stochastic processes occurring in the theory of queues and their analysis by the method of the imbedded Markov chain", *Ann. Math. Statist.*, **24**, 338-354.

Kendall, D.G. (1957) "Some problems in the theory of dams", *J. Roy. Statist. Soc.*, Ser. B, **19**, 207-212.

Khinchin, A.I. (1949) *Mathematical Foundations of Statistical Mechanics*, Dover Publications Inc., New York.

Kingman, J.F.C. (1965) "The heavy traffic approximation in the theory of queues'', Conference on Congestion Theory, Univ. of N. Carolina (to appear).

Lindley, D.V. (1952) ''The theory of queues with a single server'', *Proc. Cambridge Phil. Soc.*, **48**, 277-289.

Loève, M. (1960) *Probability Theory*, D. Van Nostrand Co. Inc., Princeton, N.J., 2nd edn.

Luchak, G. (1956) ''The solution of the single-channel queuing equations characterized by a time-dependent Poisson-distributed arrival rate and a general class of holding times'', *Operations Research*, **4**, 711-732.

Luchak, G. (1957) ''The distribution of the time required to reduce to some preassigned level a single-channel queue characterized by a time-dependent Poisson-distributed arrival rate and a general class of holding times'', *Operations Research*, **5**, 205-209.

Luchak, G. (1958) ''The continuous time solution of the equations of the single-channel queue with a general class of service-time distributions by the method of generating function'', *J. Roy. Statist. Soc.*, Ser. B, **20**, 176-181.

Lukacs, E. (1960) *Characteristic Functions*, Griffin, London.

Muskhelishvili, N.I. (1953) *Singular integral equations*, Noordhoff, Groningen.

Parzen, E. (1962) *Stochastic Processes*, Holden-Day Inc., San Francisco, Calif.

Prabhu, N.U. (1960) ''Some results for the queue with Poisson arrivals'', *J. Roy. Statist. Soc.*, Ser. B, **22**, 104-107.

Prabhu, N.U. (1964) ''Unified results and methods for queues and dams'', Conference on Congestion Theory, U. of North Carolina (to appear).

Richter, W. (1957) ''Local limit theorems for large deviations'', *Theory Prob. Applications*, **2**, No. 2, 206-220.

Rosenblatt, M. (1962) *Random Processes*, Oxford University Press.

Runnenburg, J. Th. (1960) ''Probabilistic interpretation of some formulae in queuing theory", *Bull. de l'Inst. Intern. de Statist.*, **37**, 3, 405-414.

Runnenburg, J. Th. (1964) Report S323 of the Mathematics Centre in Amsterdam.

Schwartz, L. (1950) *Théorie des Distributions*, **1**, **2**, Hermann et Cie, Paris.

Smith, W.L. (1953) "On the distribution of queuing times", *Proc. Camb. Phil. Soc.*, **49**, 449-461.

Smith, W.L. (1955a) "Regenerative stochastic processes", *Proc. Roy. Soc. (London)*, Ser. A, **232**, 6.

Smith, W.L. (1955b) "Extensions of a renewal theorem", *Proc. Camb. Phil. Soc.*, **51**, 629-638.

Smith, W.L. (1958) "Renewal theory and its ramifications", *J. Roy. Statist. Soc.*, Ser. B, **20**, 243-302.

Smith, W.L. (1962) "On necessary and sufficient conditions for the convergence of the renewal density", *Trans. Amer. Math. Soc.*, **104**, 79-100.

Spitzer, F. (1956) "A combinatorial lemma and its application to probability theory", *Trans. Amer. Math. Soc.*, **82**, 323-339.

Spitzer, F. (1960) "A Tauberian theorem and its probability interpretation", *Trans. Amer. Math. Soc.*, **94**, 150-169.

Spitzer, F. (1964) *Principles of Random Walk*, Van Nostrand Inc., Princeton, N.J.

Takács, L. (1955) "Investigation of waiting time problems by reduction to Markov processes", *Acta Math. Acad. Sci. Hungary*, **6**, 101-129.

Takács, L. (1961) "The probability law of the busy period for two types of queuing processes", *Operations Research*, **9**, 402-450.

Titchmarsh, E.C. (1948) *Introduction to the Theory of Fourier Integrals*, Oxford University Press.

Wald, A. (1945) "Some generalizations of the theory of cumulative sums of random variables", *Ann. Math. Statist.*, **16**, 287-293.

Widder, D.V. (1946) *The Laplace Transform*, Princeton University Press, Princeton, N.J.

Williams, E.J. (1961) "Conjugate distributions and saddlepoint approximations", C.S.I.R.O., Division of Mathematical Statistics.

SUPPLEMENTARY NOTES

Note 1 (page 23)

The general jump process $X(t)$ in continuous time is governed by a state-dependent transition frequency $\nu(x')$ and conditional transition distribution $A(x',x)$. When $\nu(x')$ is bounded over the state space X of the process, $\nu(x')$ and $A(x',x)$ may be replaced by a constant frequency $\nu^* = \max_{x' \in X} \nu(x')$ and $A^*(x',x) = \{1 - \nu(x')/\nu^*\} U(x-x') + \{\nu(x')/\nu^*\} A(x',x)$ without altering the process. A discussion for the lattice case is given in Section II.5. For all jump processes considered in this book, $\nu(x')$ is bounded.

Note 2 (page 41)

Specifically, the process $X_m(t)$ converges in law to the limiting process $X(t)$, i.e., all of the joint multivariate distributions $F_m(x_1, t_1; x_2, t_2; \ldots) = P\{X(t_1) \leqslant x_1 ; X(t_2) \leqslant x_2$, etc.$\}$ for $X_m(t)$ converge to those of $X(t)$.

Note 3 (page 54)

Note from (2.2) and (2.4) that $(d^2/dV^2) \log \phi_0(iV) = \sigma_V^2$. Consequently we may write $\phi_0(iV) = \exp\{\psi_0(iV)\}$, where $\psi_0(iV) = \log \phi_0(iV)$ has a second derivative which is positive in the interval of convergence for non-degenerate distributions. The function $\psi_0(iV)$ is then strictly convex, and the characteristic function $\phi_0(iV)$ is strictly superconvex.

Note 4 (pages 60, 63, 67)

Theorem III.5.1. Under the conditions (C_1) on page 60,

$$f_0^{(N)}(x) \sim \phi_0^N(iV_0)\, e^{V_0 x} (2\pi N \sigma_{V_0}^2)^{-\frac{1}{2}}, \qquad \text{as } N \to \infty. \tag{1}$$

This follows directly from the conjugate transformation $f_0^{(N)}(x) =$

$\phi_0^N(iV_0)e^{V_0 x} f_{V_0}^{(N)}(x)$ and the local form of the Central Limit Theorem (Gnedenko and Kolmogorov, 1954). This theorem, which was shown on page 60 to be applicable to $f_{V_0}(x)$ under conditions (C_1), states that $(N\sigma_{V_0}^2)^{\frac{1}{2}} f_{V_0}^{(N)}(N^{\frac{1}{2}}\sigma_{V_0} y) \to (2\pi)^{-\frac{1}{2}} \exp\{-y^2/2\}$ uniformly in y on the entire set of values $-\infty < y < \infty$. Consequently, for the sequence $y_N = N^{-\frac{1}{2}}\sigma_{V_0}^{-1} x$ we have $(N\sigma_{V_0}^2)^{\frac{1}{2}} f_{V_0}^{(N)}(x) \to (2\pi)^{-\frac{1}{2}}$, and (1) follows at once.

Theorem III.5.2. If $f_0^{(m)}(x)$ is bounded for some m, and V is any value in the interior of the convergence interval for $f_0(x)$, then

$$(2\pi N)^{\frac{1}{2}} \sigma_V f_V^{(N)}(N\mu_V) \sim 1 + \frac{Q_1(V)}{N} + \frac{Q_2(V)}{N^2} \cdots \qquad \text{as } N \to \infty, \tag{2}$$

where the expression on the right is the asymptotic series of (III.5.16). Over any closed interval I: $[V_1, V_2]$ in the interior of the convergence interval, the asymptotic series holds uniformly, i.e., for every $K \geqslant 0$,

$$N^K \left\{ (2\pi N)^{\frac{1}{2}} \sigma_V f_V^{(N)}(N\mu_V) - \sum_0^K \frac{Q_k(V)}{N^k} \right\} \to 0 \quad \text{as } N \to \infty, \text{ uniformly on } I.$$

The proof is as follows. It has been shown in Section III.4.1, under (C_1), that $\{f_V^{(m)}(x)\}^{1+\delta_V}$ is integrable $(-\infty, \infty)$ for some $\delta_V > 0$. Since I is in the interior of the convergence interval, there will be some $\delta > 0$ independent of V such that $V(1+\delta)$ is in the convergence interval for all V in I, and $f_V^{(m)}(x) \in L_{1+\delta}$ for all such V. Then, as shown under Lemma III.3.2, $|\phi_V(U)|^q \in L_1$ for some positive integer q independent of V. For $N \geqslant q$, we may therefore write, using (II.6.5),

$$f_V^{(N)}(N\mu_V) = (2\pi)^{-1} \int_{-\infty}^{\infty} \phi_V^N(U) \exp\{-iN\mu_V U\} dU \quad \text{and}$$

$$\left| 2\pi f_V^{(N)}(N\mu_V) - \int_{-a}^{a} \phi_V^N(U) \exp\{-iN\mu_V U\} dU \right| \leqslant \left\{ \int_{-\infty}^{-a} + \int_a^{\infty} \right\} |\phi_V(U)|^N dU, \tag{3}$$

for any $a > 0$. Since $\phi(U)$ is bounded away from 1 in any finite interval excluding $U = 0$, and $\phi(U) \to 0$ as $|U| \to \infty$ (Riemann-Lebesgue), the right-hand side of (3) will be less than $\rho_{aV}^{N-q} \int_{-\infty}^{\infty} |\phi_V(U)|^q dU = S_V \rho_{aV}^N$ where $\rho_{aV} = \max|\phi_V(U)|$ for $|U| \geqslant a$. The parameters S_V and ρ_{aV} are continuous in V, on the closed interval I. Consequently $S_V \leqslant S$ and, for $a > 0$, $\rho_{aV} \leqslant \rho_a < 1$. The value a will be chosen as follows.

Let d_V be the distance of the nearest zero of the analytic c.f. $\phi_V(z)$ from the origin, and e_V that of the nearest singularity. Then $\min_I d_V > 0$ and $\min_I e_V > 0$ from continuity on I. Hence there will be some circle C about $U = 0$ of positive radius in which $\phi_V(z)$ will neither vanish nor be singular for any V in I. Consider the functions $Y(z,V) = [-2\{\log\phi_V(z) - i\mu_V z\}]^{\frac{1}{2}}$, with the branch specified by requiring that $Y(z,V) \sim \sigma_V z$ as $z \to 0$. The functions $Y(z,V)$ are analytic in C and vanish only at $z = 0$. Since $Y'(z,V) = (d/dz)Y(z,V) \to \sigma_V$ as $z \to 0$ and σ_V has a positive minimum on I, there will be some circle D of radius $R > 0$ in the complex y-plane about the point $y = 0$ in which all of the functions $\Psi(z,V) = \{Y'(z,V)\}^{-1}$ are analytic and uniformly bounded over I, i.e. for which $|\Psi(z,V)| < M$. Moreover there will be some positive value $z = a > 0$ (sufficiently small) for which $Y(a,V) \in D$ for all V in I.

We may now write

$$f_V^{(N)}(N\mu_V) - \theta_1 S \rho_a^N = \frac{1}{2\pi}\int_{-a}^{a} \{\phi_V(U)\exp(-i\mu_V U)\}^N dU$$

$$= \frac{1}{2\pi}\int_{Y(-a,V)}^{Y(a,V)} \exp\{-\tfrac{1}{2}Ny^2\}\psi(y,V)\,dy\,, \qquad (4)$$

where $0 \leqslant |\theta_1| \leqslant 1$, $y = Y(U,V)$, and $\psi(y,V) = \Psi(z,V)$. Consider the complex integral $A(V) = (2\pi)^{-1}\int_0^{Y(a,V)} \exp\{-\frac{1}{2}Ny^2\}\psi(y,V)\,dy$ whose path of integration lies wholely in D. We note that $\mathrm{Re}\{Y^2(a,V)\} = -2\log|\phi_V(a)|$, and that $0 < |\phi_V(a)| < 1$. The continuity of $\phi_V(a)$ then implies that $\mathrm{Re}\{Y^2(a,V)\} > \alpha^2 > 0$ where $\alpha^2 = -2\log\max_I|\phi_V(a)|$. We may now evaluate the integral along the contour consisting of the real segment from 0 to α and the straight line from α to $Y(a,V)$. The latter will lie wholely in D, at any of its points $\mathrm{Re}\{Y^2\} > \alpha$, and its length will be less than R. Hence we have

$$A(V) = (2\pi)^{-1}\int_0^{\alpha} \exp\{-\tfrac{1}{2}Ny^2\}\psi(y,V)\,dy + \theta_2 MR\exp\{-\tfrac{1}{2}N\alpha^2\}$$

with $0 \leqslant |\theta_2| < 1$. The classical methods employed for proving

Watson's Lemma may now be applied to the integral from 0 to α. From Taylor's Theorem with the Lagrange form of the remainder,

$$\psi(y,V) = \sum_0^{2K} \frac{\psi^{(n)}(0,V)}{n!} y^n + \beta_K(V) \frac{y^{2K+1}}{(2K+1)!}$$

where the coefficients $\psi^{(n)}(0,V)$ are continuous and uniformly bounded over I, and $\beta_K(V)$ is the value of $\psi^{(2K+1)}(y,V)$ at some value y on $[0,\alpha]$. From the continuity of $\psi^{(n)}(y,V)$ in y and V, $\max|\psi^{(n+1)}(y,V)|$ over $[0,\alpha]$ is continuous in V, and $|\beta_K(V)|$ is also uniformly bounded over I. Then

$$\int_0^\alpha \exp\{-\tfrac{1}{2}Ny^2\}\psi(y,V)\,dy = \left[\sum_0^{2K} \frac{\psi^{(n)}(0,V)}{n!} \int_0^\infty \exp\{-\tfrac{1}{2}Ny^2\}y^n\,dy\right]$$

$$-\left[\sum_0^{2K} \frac{\psi^{(n)}(0,V)}{n!} \int_\alpha^\infty \exp\{-\tfrac{1}{2}Ny^2\}y^n\,dy\right]$$

$$+\left[\frac{\beta_K(V)}{(2K+1)!} \int_0^\alpha \exp\{-\tfrac{1}{2}Ny^2\}y^{2K+1}\,dy\right]. \tag{5}$$

The first bracket on the right, together with the corresponding term for the integral from $-\alpha$ to 0, gives the terms of the asymptotic series (2) to the power N^{-K} as in Section III.5.2. For the second bracket on the right, denoted by $I_2(V)$, we have $|I_2(V)| < C_2N^{-1}\exp\{-\tfrac{1}{2}N\alpha^2\}$, and for the third bracket $I_3(V)$, we have $|I_3(V)| < C_3N^{-K-1}$, where C_2 and C_3 are independent of V. Combining our results, we find that $N^{K+\frac{1}{2}}|(2\pi N)^{\frac{1}{2}}\sigma_V f_V^{(N)}(N\mu_V) - \sum_0^K N^{-k}Q_k(V)|$ is bounded over I, thus proving the theorem.

As a direct consequence of the uniformity of (2) we have the following corollary establishing (III.4.4).

Corollary III.5.2. Let $f_0^{(m)}(x)$ be bounded for some m, and let V be any value in the interior of the convergence interval of $f_0(x)$. If V_N is any sequence of values in the convergence interval, for which $V_N \to V$, as $N \to \infty$, then

$$f_{V_N}^{(N)}(N\mu_{V_N}) \sim (2\pi N\sigma_V^2)^{-\frac{1}{2}} \quad \text{as } N \to \infty. \tag{6}$$

We next present an asymptotic relation, justifying the central limit approximation (III.5.1), due to Richter (1957).

Theorem III.5.3. Let $f_0(x)$ be a probability density with characteristic function $\phi_0(z)$ such that (A) $f_0^{(m)}(x)$ is bounded for some m; (B) $\phi_0(z)$ has a convergence strip of finite width containing the real line $V = 0$ in its interior. Let x_N be any sequence of numbers for which

$$\zeta_N = \frac{x_N - N\mu_0}{N} = o(N^{-1/3}). \tag{7}$$

Then

$$f_0^{(N)}(x_N) \sim (2\pi N\sigma_0^2)^{-\frac{1}{2}} \exp[-(x_N - N\mu_0)^2/2N\sigma_0^2] \quad \text{as } N \to \infty. \tag{8}$$

For, when N is sufficiently large, the saddlepoint equation (III.3.2) will have a solution in the convergence interval. By conjugate transformation, $f_0^{(N)}(x_N) = \exp\{\theta_N x_N + N \log \phi_0(i\theta_N)\} f_{\theta_N}^{(N)}(x_N)$ where $i\theta_N$ is the saddlepoint for (x_N/N). We note from (III.5.5) that $\theta_N \sim -\zeta_N/\sigma_0^2$ as $N \to \infty$. From (7) and $x_N = N\mu_{\theta_N}$, we have $\mu_{\theta_N} \to \mu_0$, whence from (6) $f_{\theta_N}^{(N)}(x_N) \sim (2\pi N\sigma_0^2)^{-\frac{1}{2}}$. Moreover $\exp\{\theta_N x_N + N \log \phi_0(i\theta_N)\} = \exp[N\{(\mu_0 + \zeta_N)\theta_N + \log \phi_0(i\theta_N)\}] = \exp\{N(\zeta_N\theta_N + \frac{1}{2}\theta_N^2\sigma_0^2 + C_3\theta_N^3 \ldots)\}$. From (7) this is asymptotic to $\exp\{N(\zeta_N\theta_N + \frac{1}{2}\theta_N^2\sigma_0^2)\} \sim \exp\{-\frac{1}{2}N\zeta_N^2/\sigma_0^2\}$. We then have (8), as required.

More generally, for $\zeta_N = (x_N - N\mu_V)/N = o(N^{-1/3})$,

$$f_0^{(N)}(x_N) \sim \phi_0^N(iV) e^{Vx_N} (2\pi N\sigma_V^2)^{-\frac{1}{2}} \exp[-(x_N - N\mu_V)^2/2N\sigma_V^2] \quad \text{as } N \to \infty, \tag{9}$$

which follows from (8) and conjugate transformation.

Related results for distributions have been given by Cramer (1938). A variety of related results for densities have been given by Richter (1957).

Note 5 (page 68)

The reader will observe that (5.16) is an asymptotic series for $f^{(N)}(Ny)$ in powers of N^{-1} as $N \to \infty$. When y is fixed the saddlepoint is fixed, since the saddlepoint equation is $\mu_\theta = y$. When one wishes to discuss $f^{(N)}(x)$ for x fixed, the saddlepoint $i\theta$ is a function of x/N, and hence (5.16) is not an asymptotic series in powers of $1/N$. Nevertheless, the representations provided by partial sums of (5.16) will be asymptotic for $f^{(N)}(x)$, x fixed, as $N \to \infty$, provided that $\phi_0(iV)$ has a minimum at V_0 in the interior of the convergence strip (compare the discussion below equation 4.5a). The asymptotic series for $f^{(N)}(Ny)$ then provides a useful correction term for $f^{(N)}(x)$.

It should also be noted that the series (5.14) obtained from (5.13) does not require the existence of a convergence strip. If $f(x)$ has only a finite number of moments ($\geqslant 2$), the series (5.14) will have a finite number of terms, of value as correction terms to $f^{(N)}(0)$. An extensive discussion of correction terms to the Central Limit Theorem may be found in Gnedenko and Kolmogorov, Chapter 8.

Note 6 (page 74)

Equation (III.7.12) is based on the following theorem.

Theorem III.7.3. Consider the homogeneous process in continuous time $X(t) = X_J(t) + vt + X_D(t)$ of (III.7.1) for which (A) the increment distribution $A(x)$ governing $X_J(t)$ has a convergence strip containing the value V in its interior; (B) the diffusion component $X_D(t)$ is present, i.e., $D > 0$. Let x_t be any function of t for which $\zeta(t) = t^{-1}\{x_t - \mu_V(t)\} = o(t^{1/3})$, where $\mu_V(t)$ is given by (III.7.5). Then for the Green density $g_0(x,t)$ of $X(t)$, we have the asymptotic relation

$$g_0(x_t,t) \sim e^{Vx_t}\gamma_0(iV,t)\frac{\exp[-\{x_t - \mu_V(t)\}^2/2\sigma_V^2(t)]}{\sqrt{2\pi\sigma_V^2(t)}} \tag{1}$$

where $\gamma_0(z,t) = \exp[-\nu t\{1 - \alpha(z)\} - Dz^2t + izvt]$.

For by conjugate transformation, we have $g_0(x,t) = e^{Vx}\gamma_0(iV,t) \times g_V(x,t)$, where $g_V(x,t)$ is the conjugate density with parameters given by (III.7.4). Its characteristic function $\gamma_V(z,t)$ may be written as

$$\gamma_V(z,t) = \exp\left[\nu_V t\left\{-\frac{D}{\nu_V}z^2 + i\frac{v_V}{\nu_V}z - 1 + \alpha_V(z)\right\}\right] = \{\phi_V(z)\}^{\nu_V t} \tag{2}$$

where $\phi_V(z)$ is the c.f. of a probability density satisfying all the conditions of Theorem III.5.3 of Note 4, with $N = \nu_V t$, and (1) then follows from equation (8) of that note.

The approximation (III.7.12) will be valid when

$$\text{(a)} \quad \nu_V t \gg 1 \qquad \text{(b)} \quad \frac{x - \mu_V(t)}{(\nu_V t)^{2/3}\sigma_V(\nu_V^{-1})} \ll 1, \tag{3}$$

provided that the increment distribution $A(x)$ is not too far from the normal distribution. The mass point of $G_V(x,t)$ at $x = v_V t$ has the value $e^{-\nu_V t}$ at time t and will be small when (3a) is satisfied.

Note 7 (pages 115, 124)

Theorem. If $A(x)$ is absolutely continuous, and if the series $\sum_1^\infty \{A^{(n)}(x) - A^{(n)}(0)\}$ defining $H(x)$ converges, then $H(x)$ is absolutely continuous, and $\sum_1^\infty a^{(n)}(x) = h(x)$ converges almost everywhere.

Proof. Let $S^{(n)}(x) = \sum_1^n a^{(k)}(x)$. Then since $H(x)$ converges, $\lim_{n\to\infty} \int_0^x S^{(n)}(y)\,dy < \infty$. But $S^{(n)}(y)$ is non-negative and non-decreasing with n. Hence by monotone convergence, we have

$$\int_0^x \lim_{n\to\infty} S^{(n)}(y)\,dy = \lim_{n\to\infty} \int_0^x S^{(n)}(y)\,dy = H(x) < \infty .$$

Because $S^{(n)}(y)$ is non-decreasing with n, it has a limit, possibly infinite, as $n \to \infty$. The limit of $S^{(n)}(y)$ is integrable $(0,x)$, i.e., $H(x)$ is absolutely continuous and the set of values on which $h(x)$ is infinite has measure zero.

Note 8 (page 120)

The proof of Theorem V.6.1 when $A(x)$ lacks a second moment may be completed as follows.

Lemma. Let $A(x)$ and $\alpha(z)$ be the distribution and characteristic function of a random variable X. A sufficient condition for the integrability of $\{\alpha(U) - 1 - \alpha'(0)U\}/U^2$ in an interval $(-\delta,\delta)$ about zero is that

$$\int_{-\infty}^{\infty} |x| \log(1 + |x|)\,dA(x) < \infty . \tag{1}$$

If X is non-negative, the condition is both necessary and sufficient.

Proof. Let $A_1(x) = U(x) - A(x)$, where $U(x)$ is the unit step-function about zero, so that (a) $A_1(x) \leqslant 0$ and monotonic decreasing in $(-\infty,0)$; (b) $A_1(x) \geqslant 0$ and monotonic decreasing in $(0,\infty)$; (c) $A_1(-\infty) = A_1(\infty) = 0$. Because of (1), $A(x)$ has a finite first moment and therefore $A_1(x)$ is integrable† over $(-\infty,\infty)$. An integration by parts gives

$$\phi(U) = \int_{-\infty}^{\infty} A_1(x) \left\{ \frac{1 - e^{iUx}}{U} \right\} dx = i \left\{ \frac{\alpha(U) - 1 - \alpha'(0)\,U}{U^2} \right\}, \qquad (2)$$

the end terms at $\pm\infty$ vanishing† due to the finite first moment. Let $\phi(U) = C(U) - iS(U)$, where $C(U)$ and $S(U)$ are real. From (2),

$$C(U) = \int_0^{\infty} \{A_1(x) + A_1(-x)\} \frac{1 - \cos Ux}{U}\, dx;$$

$$S(U) = \int_0^{\infty} \{A_1(x) - A_1(-x)\} \frac{\sin Ux}{U}\, dx. \qquad (3)$$

Since $C(U)$ is odd in U and $S(U)$ even in U, we need only consider integrability over $(0,\delta)$. The function $B(x) = A_1(x) - A_1(-x)$ is positive, integrable, and monotonic decreasing. Because of the latter property‡, $S(U) \geqslant 0$ for all $U > 0$, but may become infinite as $U \to 0$. The integrability of $S(U)$ on $(0,\delta)$ may be seen as follows. We may write $S_T(U) = \int_0^T B(x)\, U^{-1} \sin(Ux)\, dx$ and $S(U) = \lim_{T\to\infty} S_T(U)$. Then, since $x B(x) \in L_1(0,T)$ and $U^{-1}\sin(Ux) \leqslant x$, Fubini's theorem permits us to write

$$\int_0^{\delta} S_T(U)\, dU = \int_0^T dx\; B(x) \int_0^{\delta} U^{-1} \sin(Ux)\, dU$$

$$= \int_0^T dx\; B(x) \int_0^{\delta x} w^{-1} \sin w\, dw. \qquad (4)$$

† To verify that the end terms vanish, one employs typically the standard device $x A_1(x) = x\int_x^{\infty} dA(s) \leqslant \int_x^{\infty} s\, dA(s)$ when $x > 0$.

‡ For $\int_0^T B(x) \sin Ux\, dx = \int_0^{\infty} B_T(x) \sin Ux\, dx = \sum_0^{\infty} \int_{n\pi}^{(n+1)\pi} B_T(x) \sin Ux\, dx$, and this is an alternating series of decreasing terms since $B_T(x)$ is monotonic.

But $0 \leqslant \int_0^{\delta x} w^{-1} \sin w \, dw < \pi/2$, and hence $0 \leqslant \int_0^{\delta} S_T(U)\, dU < \pi/2 \times \int_0^{\infty} dx\, B(x)$. Since $S_T(U) \geqslant 0$, it follows from Fatou's lemma that $S(U)$ is integrable on $(0,\delta)$.

For the integrability of $C(U)$ we note that $|C(U)| \leqslant \int_0^{\infty} B(x)\{1 - \cos Ux\}/U \, dx = C^*(U)$, and $C^*(U) \geqslant 0$. The integrability of $C^*(U)$ is treated in the same manner as that of $S(U)$. Since $0 \leqslant \int_0^{\delta x} w^{-1}(1 - \cos w)\, dw < K + \log(1+x)$ for some $K > 0$, it follows as above that $C(U)$ and $C^*(U)$ are integrable on $(0,\delta)$ if $B(x)\log(1+x) \in L_1(0,\delta)$, as well. Finally one notes that $B(x)\log(1+x) \in L_1(0,\infty)$ when $\int_0^{\infty} x \log(1+x)\, dB(x) < \infty$, and this is implied by (1).

When the random variable X is non-negative, $C(U) = C^*(U) \geqslant 0$ and the necessity of condition (1) is apparent. Q.E.D.

The integrability of $\{\alpha(U) - \beta(U)\}/U^2$ required for Theorem V.6.1 follows.

When $A(x)$ has a non-zero first moment μ, the extended renewal function $H(x) = \sum_1^{\infty}\{A^{(k)}(x) - A^{(k)}(0)\}$ converges for all x. An analytical proof of this, similar to that given by Spitzer (1964), p. 86, for the lattice case, may be based on a theorem of Chung and Fuchs (1951). One may also demonstrate analytically, as in Feller and Orey (1961), that for $\mu > 0$, $\{H(x+a) - H(x)\} \to \mu^{-1}a$ as $x \to \infty$ (Blackwell's Theorem) and $\{H(x+a) - H(x)\} \to 0$ as $x \to -\infty$.

A local result due to Smith (1955b, 1962) states that if (a) $f(x) \in L_{1+\delta}$ for some $0 < \delta \leqslant 1$; (b) $f(x) \to 0$ as $x \to \pm\infty$; (c) $f(x)$ has a first moment $\mu > 0$; then $h(x) \to \mu^{-1}$ at $+\infty$ and $h(x) \to 0$ at $-\infty$. Smith's result may be used to weaken condition (3) of Theorem V.6.1 as follows:

Theorem V.6.1′. Condition (3) of Theorem V.6.1 may be replaced by (3′) $x\,a(x) \in L_1$.

Proof. Since $a(x)$ is bounded, $a(x)$ and $\alpha(U)$ belong to $L_2(-\infty,\infty)$ and $a^{(2)}(x)$ is bounded and continuous for all x. Let $h_2(x) = \sum_1^{\infty} a^{(2k)}(x)$. Then $h_2(x) = a^{(2)}(x) + \int_{-\infty}^{\infty} a^{(2)}(x-x')h_2(x')\, dx'$. From the asymptotic convergence of $h_2(x)$ we have, for some X_1, X_2, A_1, A_2, that $0 \leqslant h_2(x) < A_1$ for $x < X_1$, and $0 \leqslant h_2(x) < A_2$ for $X > X_2$. Then by dividing the interval of integration $(-\infty,\infty)$ into the intervals $(-\infty,X_1)$, (X_1,X_2), and (X_2,∞) we see that $h_2(x) \leqslant a^{(2)}_{max} + a^{(2)}_{max}[H_2(X_2)-H_2(X_1)]$

$+ A_1 + A_2$, i.e., $h_2(x)$ is bounded. Consequently† we have that $\int_{-\infty}^{\infty} a^{(2)}(x-x')h_2(x')\,dx'$ is continuous for all x and hence $h_2(x)$ is as well. We have, moreover, $h(x) - a(x) = h_2(x) + \int_{-\infty}^{\infty} h_2(x-x')a(x')\,dx'$, and the latter term is also continuous for all x by dominated convergence. The remainder is straightforward.

An earlier version of Theorem V.6.1′ (Keilson, 1964b) tacitly assumed a certain integrability without justification, as pointed out to the author by W.L. Smith in a private communication.

To complete the proof of Theorem 6.2 when $A(x)$ does not have a second moment, we note from the lemma above that $x\,a(x)\log(1+|x|) \in L_1$ implies that $\{\alpha(u)\exp(-i\mu_1 U) - 1\}/U^2 \in L_1(-\delta,\delta)$. Hence $\{\alpha^N(U) \times \exp(-iN\mu_1 U) - 1\}/U^2 \in L_1(-\delta,\delta)$, and this is all that is required for the theorem to go through. It is also clear from Theorem V.6.1′ above that the availability of a first moment is sufficient.

Note 9 (page 127)

Lemma. If, for the set of lattice probabilities e_n, $\sum_{-\infty}^{\infty} |n| \log(1+|n|)\, e_n < \infty$, then $|\{\epsilon(u) - 1 - \epsilon'(1)(u-1)\}/(u-1)^2|$ is integrable over the unit circle.

This may be seen directly from the lemma of Note 8 and the transformation $u = e^{i\theta}$, $\epsilon(u) = \alpha(\theta)$.

To complete the proof of Theorem V.7.1, let $\epsilon(u)$ be the g.f. of a lattice random variable with mean $\mu \neq 0$ and let $\phi(u)$ be the rational generating function of another such variable with the same mean, i.e., $\epsilon'(1) = \phi'(1)$. Let $h_n^{\epsilon}(w) = \sum_{k=1}^{\infty} w^k e_n^{(k)}$ and $h_n^{\phi}(w) = \sum_{1}^{\infty} w^k f_n^{(k)}$, with $0 \leqslant |w| < 1$. Then $\gamma_n(w) = h_n^{\epsilon}(w) - h_n^{\phi}(w)$ is given by

$$\gamma_n(w) = \frac{1}{2\pi i} \oint_{\substack{\text{unit}\\ \text{circle}}} \frac{w\{\epsilon(u) - \phi(u)\}}{\{1 - w\epsilon(u)\}\{1 - w\phi(u)\}} \frac{du}{u^{n+1}}$$

By the lemma above, the g.f. of $\gamma_n(w)$ is absolutely integrable and $\gamma_n(w)$ bounded for all $0 \leqslant w \leqslant 1$. Moreover as in Section V.6, $\gamma_n(w) \to \gamma_n(1)$ as $|w| \to 1-$ by dominated convergence. Since $\gamma_n(1) \to 0$ as

---

† If $f(x) \in L_1(-\infty,\infty)$, then $\int_{-\infty}^{\infty} |f(x+\theta) - f(x)|\,dx \to 0$ when $\theta \to 0$. See for example N. Wiener, *The Fourier Integral*, Cambridge, 1933, p. 14.

$n \to \pm\infty$ by the Riemann-Lebesgue lemma, the proof of Theorem V.7.1 is complete.

Note 10 (page 146)

For the bounded process $X_0 = 0$, $X_k = \max[0, X_{k-1} + \xi_k]$ of Section V.1, an important relation known as Spitzer's Identity is available between the distributions $F_k(x)$ and $A^{(k)}(x)$. This identity, originally obtained by Spitzer (1956) by combinatorial reasoning, may be obtained from (V.12.1) and Wiener-Hopf methods in the following way. Since $X_0 = 0$, and $\phi_0^+(z) = 1$, (V.12.1) has the form of a homogeneous Hilbert problem when $|w| < 1$. Consequently,

$$\log\left\{\frac{\phi^+(w,z)}{\phi^+(w,\infty)}\right\} = -[\log\{1 - w\,\alpha(z)\}]^* = \sum_1^\infty \frac{w^n}{n}[\alpha^n(z)]^*$$

when $A(x)$ is absolutely continuous, so that $\alpha(\infty) = 0$. The asterisk denotes the Cauchy integrals for the upper half-plane associated with the Hilbert problem. Hence we find

$$\phi^+(w,z) = \phi^+(w,\infty)\exp\left\{\sum_1^\infty \frac{w^n}{n}\int_{0+}^\infty e^{izx}\,dA^{(n)}(x)\right\}. \tag{1}$$

But $\phi^+(w,0) = \sum_0^\infty w^k = (1-w)^{-1} = \exp\{\sum_1^\infty \frac{w^n}{n}\}$, and $\int_{0+}^\infty dA^{(n)}(x) = P\{S_n > 0\}$, where $S_n = \sum_1^n \xi_i$. We thus obtain

$$\sum_0^\infty w^k \int_{-\infty}^\infty e^{izx}\,dF_k(x) = \exp\left\{\sum_1^\infty \frac{w^n}{n} P\{S_n \leqslant 0\} + \sum_1^\infty \frac{w^n}{n}\int_{0+}^\infty e^{izx}\,dA^{(n)}(x)\right\} \tag{2}$$

and this is *Spitzer's Identity*. The requirement that $A(x)$ be absolutely continuous is easily removed. The characteristic function of the ergodic distribution is given by $\phi_\infty(z) = \lim_{w\to 1-}(1-w)\phi(w,z)$. Hence from (1) we have

$$\phi_\infty(z) = \exp\left\{\sum_1^\infty n^{-1}\int_{0+}^\infty (e^{izx} - 1)\,dA^{(n)}(x)\right\}. \tag{3}$$

By combinatorial reasoning, Spitzer (1956) also obtained the now classical results of Sparre Andersen, and extended them. These results are also available from a Wiener-Hopf analysis of the passage process on the non-positive half-line for the homogeneous process $X_k = \sum_1^k \xi_i$. For let $f_k(x)\,dx$ be the probability that X_k has not been positive and is in $(x, x+dx)$. Then $f_0(x) = \delta(x+0)$ and $f_n(x) = [f_{n-1}(x) \cdot a(x)]\,U(-x)$, where the dot denotes spatial convolution, $U(x)$ is the unit step function, and $a(x)$ is the increment density (absolute continuity again being assumed for simplicity). For $f(w,x) = \sum_0^\infty f_n(x) w^n$ we have $f(w,x) = \delta(x) + w[f(w,x) \cdot a(x)]\,U(-x) = \delta(x) + w[f(w,x) \cdot a(x)] - r(w,x)$ where $r(w,x) = w[f(w,x) \cdot a(x)]\,U(x)$. Fourier transformation gives

$$\phi^-(w,z)[1 - w\,\alpha(z)] = 1 - \rho^+(w,z) \tag{4}$$

and again one has a standard homogeneous Wiener-Hopf problem. As for (2), we then have

$$\rho^+(w,z) = 1 - \exp\left\{-\sum_1^\infty \frac{w^n}{n} \int_0^\infty a^{(n)}(x)\,e^{izx}\,dx\right\}. \tag{5}$$

But $r(w,x) = \sum_1^\infty w^k Q_k b_k(x)$ where Q_k is the probability that S_k has its first positive value at time k and $b_k(x)$ is the probability density for the positive value then assumed. Equation (5) is the basic result of Spitzer. If we set $z = 0$ in (5) we obtain

$$\sum_1^\infty w^k Q_k = 1 - \exp\left[-\sum_1^\infty w^k P\{S_k > 0\}\right] \tag{6}$$

which is Sparre Andersen's result. A simple derivation of (5) has also been given by Feller (1959).

Note 11 (page 167)

Theorem VI.7.2 implies, as a special case, an important relation between the maximum of the first k partial sums of a sequence of independent random variables, and the Lindley process $X_0^L = 0$, $X_k^L = \max[X_{k-1}^L + \xi_k, 0]$ of Section V.1. If we let $y = L - x$ with $x > 0$ and permit L to go to infinity, equation (7.11) states that, for all $k \geqslant 1$,

$$P\left\{\max_{1 \leqslant j \leqslant k} \sum_1^j \xi_i \geqslant x\right\} = 1 - P\{X_k^L < x\} \qquad x > 0, \tag{1}$$

and hence that

$$P\left\{\max_{1\leqslant j\leqslant k}\sum_{1}^{j}\xi_i \leqslant x\right\} = P\{X_k^L \leqslant x\} \qquad x \geqslant 0. \tag{2}$$

When the mean increment is negative, the Lindley process is ergodic, and the distribution of the largest partial sum $\sum^{j} \xi_i$ ever attained coincides with the ergodic distribution of the Lindley process. Relations similar to (2) for more general homogeneous processes are reviewed in Prabhu (1964).

It may be noted that the result of Sparre Andersen (Note 10, equation (6)) follows from (1) and Spitzer's Identity. For it is clear that the probability that a process $X_k = \sum_1^k \xi_i$ will have its first positive value at the epoch k is given by $P\{\max_{1,\ldots,k} X_j > 0\} - P\{\max_{1,\ldots,k-1} X_j > 0\}$ $= P\{X_{k-1}^L = 0\} - P\{X_k^L = 0\}$. The reader will then verify that equation (6) of Note 10 follows with the aid of the Spitzer Identity, equation (2) of that note.

Note 12 (page 124)

If in Theorem V.6.4 the random variable ξ is non-negative with $P\{\xi = 0\} = \alpha > 0$, the function $H^+(w,x) = \sum_1^\infty w^k A^{(k)}(k)$ has only one singularity in the finite w-plane located at $w = \alpha^{-1}$. For we have

$$H^+(w,x) = wA(x) + wA(x)\cdot H^+(w,x) \tag{1}$$

where the dot denotes convolution. Let $A(x) = \alpha\, U(x) + (1-\alpha)A^*(x)$, where $A^*(x)$ is a probability distribution with purely positive support, and let $H^+(w,x) = w\alpha(1-w\alpha)^{-1}U(x) + F(w,x)$. From simple algebra we see that $F(w,x)$ obeys a Volterra equation with the same structure as (1) which has the unique solution

$$F(w,x) = (1-w\alpha)^{-1}\sum_1^\infty \left\{\frac{(1-\alpha)w}{1-\alpha w}A^*(x)\right\}^{(k)}. \tag{2}$$

Hence

$$H^+(w,x) = \frac{w\alpha}{1-\alpha w}U(x) + \frac{1}{1-\alpha w}H^{*+}\left(\frac{(1-\alpha)w}{1-\alpha w}, x\right) \tag{3}$$

where $H^{*+}(w,x) = \sum_1^\infty w^k A^{*(k)}(x)$ is an entire function by Theorem V.6.4. It follows from (3) that $H^+(w,x)$ is analytic at all w other than α^{-1}. In general the singularity will be essential, but in some instances it will be a pole of finite order.

INDEX